Exploring Electricity and Electronics
basic fundamentals

by

HOWARD H. GERRISH

Professor Emeritus
Humboldt State University
Arcata, California

and

WILLIAM E. DUGGER, JR.

Professor and Program Area Leader
Virginia Polytechnic Institute
and State University
Blacksburg, Virginia

South Holland, Illinois
THE GOODHEART-WILLCOX COMPANY, INC.
Publishers

Library of Congress Cataloging in Publication Data

Gerrish, Howard H.
   Exploring electricity and electronics.

   Includes index.
   1. Electric engineering.   2. Electronics.
I. Dugger, William, joint author.   II. Title.
TK146.G442   621.3   80—20830
ISBN 0—87006—308—1

Exploring Electricity and Electronics contains the most complete and accurate
information available at the time of publication. It is your responsibility to follow
all safety procedures. The Goodheart-Willcox Company, Inc. cannot assume
responsibility for any changes, errors or omissions.

INTRODUCTION

EXPLORING ELECTRICITY AND ELECTRONICS is a first course of study in this interesting and ever changing field. This text will acquaint you with the basic fundamentals, latest technology and practical applications of both electricity and electronics.

This revised edition of the text is updated throughout and features a new Unit 11 on "Integrated Circuits." Current electronic concepts are explained. Technical terms are defined when introduced. Some of the latest electronic applications are pictured, and extra color has been added to the line drawings to make them more meaningful and easier to understand.

Many worthwhile projects have been carried over to this revised edition of EXPLORING ELECTRICITY AND ELECTRONICS, and some new projects have been added. At least one of these simple projects is included in each unit. Build as many as you can. Building the projects and performing the interesting experiments will aid you in understanding the principles of electricity and electronics.

Many of the finest technological advances of the past few decades have been made in the field of electronics. This challenging field continues to expand and develop with new discoveries and applications. Through the pages of this text, you can learn the fundamentals of electronics and explore the rewarding careers that abound in this field.

HOWARD H. GERRISH
WILLIAM E. DUGGER, JR.

CONTENTS

ELECTRONIC PROJECTS

To develop a degree of skill and understanding, the following projects are included in this text. They are listed in numerical order by page number.

An exploratory electronics laboratory, suitable for instruction as covered by this text, is available from the Lab-Volt Division, Buck Engineering Co., Inc., Box 686, Farmingdale, NJ, 07727.

Student experiment systems that include manuals, lesson plans, component parts, connecting leads and power and metering equipment are available through Lab-Volt and others.

Fig. 1-1. A modern microwave oven.   (Amana)

Fig. 1-2. Home computer system.
(Processor Technology Corp.)

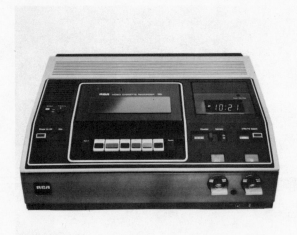

Fig. 1-3. Home video cassette recorder system.   (RCA)

Fig. 1-4. Home electronic television games.   (Atari)

Fig. 1-5. Citizens Band radio.   (Cobra)

Unit 1

THE ELECTRONIC AGE

In this first unit, you will learn:

1. How electricity is used in the home.
2. How electricity and electronics are utilized in industry.
3. What applications are made in the field of transportation.
4. How electricity and electronics are used in the communications industry.
5. Where electronics is used in the medical fields.

ELECTRICITY AT HOME

The principles of electricity and electronics you will explore in this course find wide application in the home.

You can get an idea of the scope of electrical power used in the home from the list which follows:

LIGHT
Incandescent lighting, fluorescent lighting and high intensity lights.

FOOD PREPARATION AND COOKING
Electric ranges and ovens, refrigerators, deep-freezers, ice cube makers and crushers, small appliances for mixing, blending, grinding, cutting and cooking, dishwashers and roasters, garbage disposals, can openers and microwave ovens. See Fig. 1-1.

HEAT AND AIR CONDITIONING
Electric space heating, air conditioning, ventilation, automatic controls, electronic air filtering, humidifiers and sump pumps.

PERSONAL CARE
Electric razors, toothbrushes, massagers, electric pads and blankets, vibrators, shoe polishers and brushes, hearing aids and hair dryers.

HOUSEHOLD CONVENIENCE
Radio controlled garage doors, floor polishers, intercommunications systems, photo light control, elevators, vacuum cleaners and home computers (microprocessors). See Fig. 1-2.

ENTERTAINMENT
Film and slide projectors, AM and FM radio, record changers, stereophonic sound systems, black and white TV, color TV, musical instruments, organs, guitars, recording equipment, color organs, TV recording equipment, Fig. 1-3, and television games, Fig. 1-4.

HOBBIES
Power machinery and tools, drills, sanders, paint sprayers, ham radio and CB radio, Fig. 1-5.

GARDEN MAINTENANCE
Lawn mowers, cultivators, small tools, hedge trimmers, auto washers and automatic sprinkler systems.

COMMUNICATIONS
Telephones, videophones.

SAFETY AND SECURITY
Burglar alarm systems and fire alarm systems.

The list is far from complete. It does not list all items in the various categories and does not include electric typewriters, calculators and computers found in many homes. See Fig. 1-6. Nor does it include the family car, which would remain in the garage if it were not for the storage battery.

Fig. 1-6. Programmable scientific pocket calculator. (Hewlett-Packard)

ELECTRICITY AND ELECTRONICS IN INDUSTRY

In industry, as in the home, the electric motor can well be called "servant." From a tiny motor to drive a delicate instrument to giant motors capable of developing hundreds of horsepower, the motor is ready and able to perform at the flip of a switch. It is quiet and efficient, requires little attention and is relatively easy to connect to the machine and power source. Motors provide a very economical method of converting electrical energy into mechanical energy.

The generation of electric power is also an industry in its own right. In past years, most power generators were turned by steam turbines. The dwindling supply of coal has encouraged engineers to search for other sources of power. In some areas, water is stored behind dams as a source of potential energy. As water is released, it flows through a hydroelectric plant where it turns turbine generators.

Since the first explosion of an atomic bomb, scientists have searched for ways and means of converting the enormous energy of the atom into peaceful uses. The atomic power plant is one such accomplishment. The nuclear reactor is used to heat steam to drive the turbine-generators. Atomic energy may well supply the answer to our increasing needs for electric power. Fig. 1-7 shows a large nuclear power plant and a fossil fuel electric generating plant.

Electronics in industry appears more in instrumentation and control of machinery and processes. Manufacturing processes that once required close and visual supervision may now be observed and controlled from remote locations. This alone increases safety and economy in manufacturing.

The conversion of electrical energy to heat energy will also find broad usage in industry. All kinds of heating requirements are met, from heating a small office to a high temperature electric furnace for melting steel. One less well known application of heating is inductive heating. It is used extensively in heat treating and tempering metals. The part to be heated is placed in a coil. High currents are induced in the material to be heated by transformer action, and the part is heated by its own internal molecular friction.

Electricity also is the servant of electric welding, the process in which metal parts are joined together by fusion (combining metals by melting together) of the metals. From building automobiles to battleships, the electric arc welder serves industry by increased production speed and economy.

Another common industrial use of electrical energy, which may find its way into your home, is ultrasonic cleaning and washing. In this application the parts to be washed are not sloshed around in soapy water, but are placed in a tank filled with the cleaning liquid. The liquid is caused to vibrate by an electronic generator at ultrasonic frequencies.

ELECTRICITY AND ELECTRONICS IN TRANSPORTATION

Transportation by land, by air or by sea uses electricity for power, direction and control.

Fig. 1-7. Nuclear (left) and fossil fuel (right) electricity generating plant in South Carolina. (Carolina Power and Light Co.)

Electric trains used in many areas are fast and clean, and provide services and comforts that were only dreams a few years ago. The increased railroad traffic demands precise control and scheduling. In this capacity, electronics steps forward with electronic observation and control; computers provide fast computations and decisions. These are requirements of this Electronic Age.

The compass is still a vital instrument in navigation. Today, however, electronic navigational systems are widely used in ships and aircraft. Immediate voice communication with the home port is provided by radio. Radar is used to survey the horizon for approaching ships and dangerous reefs. The ship's captain can "see in the dark." Electronic depth sounders immediately indicate the distance to the ocean floor. To aid in navigation, radio compasses can instantly show the position of ship or aircraft.

Today, many ships are powered by electricity. In addition, electricity provides the power aboard ship to operate hoists, winches, cranes and elevators. On naval ships, electronics takes over the control of "fire power." Guns are pointed, trained and fired, all by electricity and electronic circuits.

Probably the most sophisticated and complex system of electronic circuits is used to control and direct our aircraft. Fig. 1-8 shows the cockpit of the supersonic Concorde. Needless to say, extremely skilled pilots and engineers are required for its safe and efficient operation. On the ground, highly skilled and trained technicians must inspect and maintain all aircraft at peak performance. Fig. 1-9 shows a supersonic passenger plane landing.

ELECTRICITY AND ELECTRONICS IN THE COMMUNICATIONS INDUSTRY

Communicating by radio waves is an important division of electronics. At home, in in-

Fig. 1-8. Cockpit of the Concorde.   (British Airways)

Fig. 1-9. The Concorde comes in for a landing.
(British Airways)

dustry, in transportation and in space travel, communication by radio waves is a part of each system. In television, the picture and sound require separate transmitters.

In recent years, video tape recording has changed the nature of TV broadcasting. Studio programs can be taped at a convenient time, and the tape is held until the program is needed. By video tape recording, we enjoy the instant playback of athletic teams in action. It also is a rich source for summer "reruns."

Large screen television is here. Fig. 1-10 shows a family viewing a televised show on a large screen TV.

Fig. 1-10. Large screen television.   (Atari)

The telephone provides a dependable and inexpensive method of electronic communications. Great improvements are being made in the telephone industries, Fig. 1-11. Telephone lines which reach around the world use wires and microwave radio. Others bounce signals from communcations satellites. And now the Picturephone has become practical. With this, you can SEE and HEAR during your telephone conversations.

Fig. 1-11. Telephone operators in modern telecommunications center.   (United Telecommunications, Inc.)

ELECTRONICS IN MEDICINE

The science of electronics serves well in the hands of doctors and surgeons in the hospitals of the world. Many of the instruments devised provide continuous observation, monitoring and recording the condition of patients. Alarm systems are provided to call nurses and doctors in emergency cases. Information such as pulse rate, respiration, blood pressure and body temperature is available instantly.

Other instruments are designed to use electronic circuitry to aid in body functions. Such is the "Gift of Life" made possible by the Pacemaker which provides an electrical stimulus to cardiac (heart disease) patients. See Fig. 1-12.

Your study of the preceding pages will give meaning and understanding to the following Units. Electronics is an extremely interesting study and a challenge to all youths who live in the Age of Electronics.

LED TESTER PROJECT

A simple, yet effective, tester can be made from a light emitting diode (LED), a resistor and a couple of probes. See Fig. 1-13. The LED tester can be used to test low voltages, polarities and power supply output voltages. This little unit is housed in an encapsulated case which can be made by pouring plastic casting resin into a 35 mm film can.

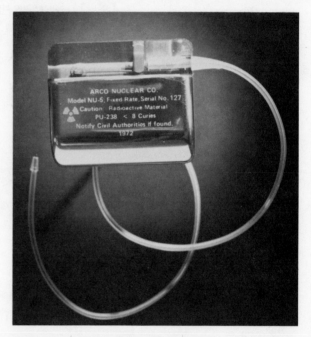

Fig. 1-12. A Nuclear Cardiac Pacemaker.   (U. S. Atomic Energy Commission)

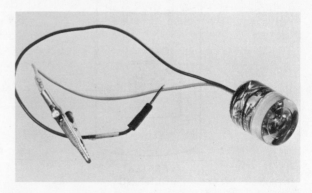

Fig. 1-13. Light Emitting Diode (LED) Tester. (Sam Popkins, Hammond Junior High School, Alexandria, VA)

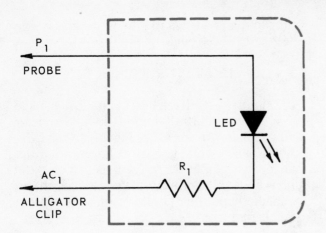

PARTS LIST FOR LED TESTER

R_1 — 220 Ω, 1/2W resistor
LED — light emitting diode, XC111C or ED 126 or equivalent
P_1 — pin plug
AC_1 — alligator clip with insulated cover
Misc. — 5/8 in. square of printed circuit board, 1 ft. red stranded test wire, 1 ft. black stranded test wire, plastic casting resin, plastic 35 mm film can (for casting mold), solder, resist, etchant (Fe_2Cl_3)

Fig. 1-14. Schematic and parts list for LED Tester.

The first step is to collect the component parts. See the schematic and parts list in Fig. 1-14. Next, fabricate the printed circuit by selecting one of the designs from Fig. 1-15.

Mount the LED and resistor on the printed circuit along with the test leads. Solder the alligator clip and pin plug onto the other ends of the test leads. Test the circuit to make sure

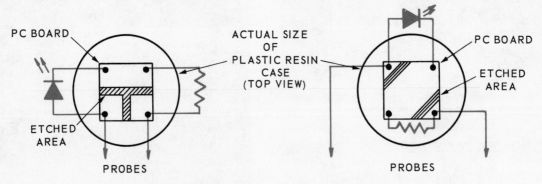

Fig. 1-15. Some printed circuit designs for LED Tester.

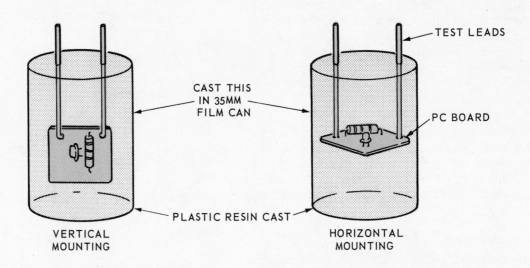

Fig. 1-16. Mounting printed circuit board in encapsulated resin.

the device works properly, then encapsulate it in the resin. Use a commercial casting resin for this process. See Fig. 1-16.

To see some of the LED tester applications, refer to Fig. 1-17.

RULES OF SAFETY

Acquire the safety habit.
Always think first and observe caution.
Keep your shop and bench clean and in order.
Use the right tool for the right job.
Operate machines only after proper instruction.
Wear goggles, face shields and protective clothing when required.
Do not horse around. No jokes.
Do not interrupt or talk to a fellow worker who is operating a machine.
Handle all materials carefully.
Know where the first aid kit is located and always get first aid for any injury.
Use common sense.

OUTSIDE ACTIVITIES

1. Plan a field trip to a local industry and report on the types of electrical machinery and equipment in use. Identify the electrical effect produced such as heat, light, sound or motion.
2. From your reading or field trip, select one particular machine or process involving electricity and make a verbal detailed explanation to your class. This will probably require some study and research.
3. Make an appointment with an electrician, technician or engineer and ask about this particular vocation or profession. Prepare the questions you wish to ask before the interview. Make a verbal report to your class.
4. Make a list of books (and their authors) on electricity and electronics in your school library. Post the list on the shop bulletin board.
5. Arrange for an electrician, technician or engineer to visit your class and talk about their vocation. Prepare questions before their arrival.
6. Count and list the purpose of all the electric motors used in your home.

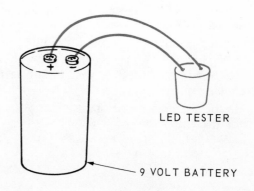

1 BATTERY TEST
AND
2 TEST POLARITIES

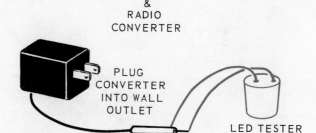

3 TEST CONVERTER OUTPUT

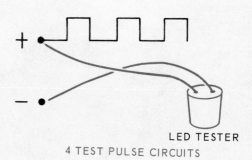

4 TEST PULSE CIRCUITS

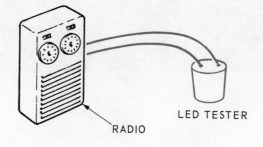

5 TEST LOW VOLTAGES IN A CIRCUIT

Fig. 1-17. Some applications for LED Tester.

7. Prepare a list and, if possible, secure samples of magazines on electricity and electronics available at your local news stand. Keep the magazines for your fellow students to look over. You may wish to subscribe to one or more.

8. Is there a ham radio operator in your class or do you know of a friend who has this hobby. Ask the ham to talk about the hobby and demonstrate the radio equipment (gear).

9. Ask a fellow student or teacher who is interested in hi-fi amplifiers to talk about and demonstrate the equipment. Also make tape recordings of your class discussions or talks by engineers and technicians.

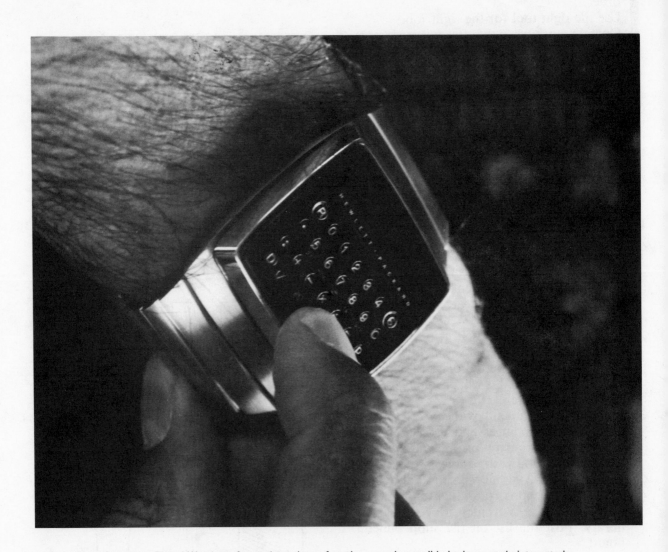

Calculator-Digital Watch performs three dozen functions, made possible by large scale integrated circuits providing equivalent of 38,000 transistors.   (Hewlett-Packard)

Unit 2

WHAT IS ELECTRICITY?

Electricity is a kind of energy. The term ENERGY may be defined as the capacity or ability to do work. ELECTRICITY is the movement of electrons in a conductor (wire or other material that will allow current to flow).

The following major subject areas are covered in this unit:

1. What potential energy is.
2. Where matter is found in this world.
3. What atoms are, and how they are made up.
4. What electrons and protons are.
5. How charged materials act if they are "like charged" or "unlike charged."
6. How an electroscope can be used to demonstrate charges.

POTENTIAL ENERGY

POTENTIAL ENERGY is energy at rest. For example, a scientist may describe a body of water, such as the mountain lake shown in Fig. 2-1, as possessing POTENTIAL ENERGY.

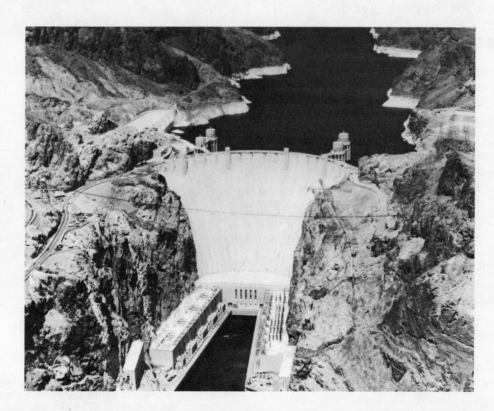

Fig. 2-1. Aerial view of Hoover Dam hydroelectric plant. (Dept. of Energy, Bureau of Reclamation)

This means the water in the lake has the capacity to do work. When released to flow into a power plant, energy from the falling water operates a turbine to generate electricity.

Why is the water falling? It falls because of gravity. The laws of gravity were formulated by Sir Isaac Newton in the 17th century.

At this point, we need to learn more about the potential energy stored in the mountain lake. How do we measure this POTENTIAL? The following example will help you understand the measurement of energy:

If you were to lift a ten pound weight to a height of three feet, the weight would acquire thirty FOOT-POUNDS of POTENTIAL ENERGY (3 ft. x 10 lb. = 30 ft. lb.). If you dropped the weight (not on your foot), the STORED ENERGY could do THIRTY FOOT-POUNDS of WORK.

POTENTIAL ENERGY can be further defined as energy stored as a result of position. Our lake has potential energy due to its height or elevation above the power plant.

OUR WORLD IS MADE OF MATTER

MATTER is defined as anything which takes up room or occupies space. Matter may be in the form of a solid, liquid or gas. All matter is made up of the basic elements of nature. Each element has been given a name. Familiar elements include gold, silver, iron, oxygen, hydrogen and carbon. There have been over one hundred elements discovered and identified.

Matter consists of the basic elements, either in pure form or in mixtures and compounds. For example: water consists of two parts of hydrogen and one part of oxygen (H_2O); salt consists of one part sodium and one part chlorine (NaC1).

ATOMS - PROTONS AND ELECTRONS

Elements may be divided into parts. The smallest unit into which an element may be divided without losing its identifying characteristics is called a MOLECULE. A molecule consists of one or more like atoms in an element or two or more different atoms in compound.

An atom is the smallest part of an element having all of the properties of the element. ATOMS consist of electrically charged particles. Some particles have a positive charge (+ charge) and are called PROTONS. Some have a negative charge (− charge) and are named ELECTRONS. Others have neither a positive nor negative charge and are called NEUTRONS. We are concerned in particular with the ELECTRON.

A MODEL OF THE ATOM

A simplified explanation of the structure of an atom is shown in Fig. 2-2. For our discussion, we may consider that the center or NUCLEUS or the ATOM is made up of PRO-

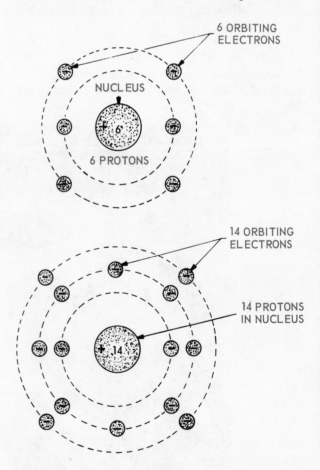

Fig. 2-2. Top. The Carbon atom has six ELECTRONS in its outer rings or orbits. Its atomic number is 6. Bottom. The Silicon atom has fourteen ELECTRONS in its outer rings or orbits. Its atomic number is 14.

TONS and NEUTRONS. Surrounding the nucleus in specific rings or orbits are found the ELECTRONS. In the illustration, the atomic models of carbon (C) and silicon (Si) are shown. Note that a carbon atom has six protons or positive charges in the nucleus and two orbits containing six negative charges or ELECTRONS. In the silicon atom, there are fourteen positive charges or protons in the nucleus and three orbits containing fourteen negative charges or ELECTRONS.

GREEDY OR GENEROUS

"Greedy" and "generous" are peculiar words to use in the study of electricity. Why are they used? Looking back to Fig. 2-2, note that in both atomic models the NUMBER OF PROTONS IS EQUAL TO THE NUMBER OF ELECTRONS. Since they are equal in number and opposite in charge, the atom has NO ELECTRICAL CHARGE. Therefore, the atom is NEUTRAL.

Study the location of the PROTONS for a moment. They are gathered together in the nucleus and are difficult to reach. However, the electrons are moving about in circles and are rather easily disturbed. If some electrons could be removed from their orbits, then the NUMBER OF PROTONS WOULD BE GREATER THAN THE NUMBER OF ELEC-

TRONS. The PROTONS would be in the majority and the atom would indicate a POSITIVE CHARGE. A positive ION has been made.

Now look at the opposite situation. Suppose an atom is greedy and would capture an electron from a closely associated but generous atom. Then the atoms would have MORE ELECTRONS THAN POSITIVE PROTONS. The ELECTRONS would be in the majority. The atom would indicate a NEGATIVE CHARGE. A negative ION has been produced.

In Fig. 2-3, the chemical reaction to form a simple grain of table salt is shown. Which atom is generous and which is greedy? Chemical changes are assumed to occur in this manner.

THE LAW OF ELECTROSTATIC CHARGES

Some interesting experiments may be performed to demonstrate the presence of an electrical charge and how these charges behave. Perform the experiments that follow. If you understand the Laws of Electrostatic Charges, you are well on your way toward an exciting study of electricity. ELECTROSTATICS means "electricity at rest." Later on you will study "electricity on the move" or current.

You probably have experienced the generation of an electrostatic charge with surprising

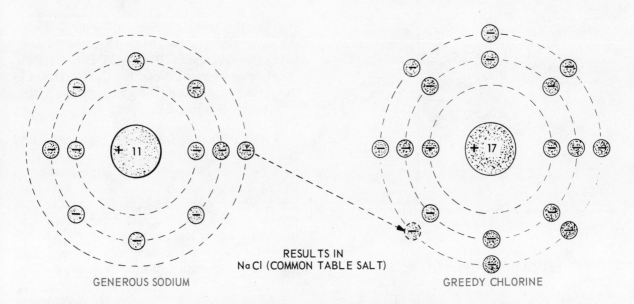

RESULTS IN
NaCl (COMMON TABLE SALT)

GENEROUS SODIUM

GREEDY CHLORINE

Fig. 2-3. The atomic structure of Sodium and Chlorine.

results. For example, if the family car has the usual type of seat covers and you slide over these covers, you may get a sudden shock when you touch the metal door handle.

Have you a family pussycat? Pat your kitty and be rewarded with a friendly purr. Now bring your hand in a backwards direction just above the kitten's fur. Note how his fur will stand on end. Sometimes you will hear the crackling of electric sparks. (Warning - this may not be appreciated by the cat.)

Take a toy balloon, blow it up and tie its end. Rub the balloon vigorously with a piece of wool cloth. Now touch the balloon to the wall of your room. It mysteriously sticks to the wall. An explanation of this mystery is given in Fig. 2-4. Friction produces a negative charge on the balloon. As the balloon approaches the wall, it repels the electrons in the wall which leaves a positive area. This demonstrates the LAW OF CHARGES:

LIKE CHARGES REPEL EACH OTHER
UNLIKE CHARGES ATTRACT EACH OTHER

The fact that one electron repels another electron is a fundamental law in electricity.

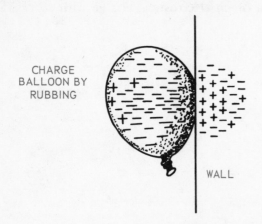

CHARGE BALLOON BY RUBBING

WALL

Fig. 2-4. A balloon is attracted because unlike charges are produced by the charged balloon.

ELECTROSCOPE PROJECT

Definite conclusions and proof of the Laws of ELECTROSTATICS may be gained by building a simple ELECTROSCOPE. This instrument detects the presence and polarity (negative or positive) of an electrostatic field. Any charged body will have an INVISIBLE FIELD OF FORCE surrounding it. This will be proved by the electroscope.

Constructional details on an electroscope are given in Fig. 2-5. Any glass jar may be used. Bend a twelve inch length of heavy copper wire to form the center rod. On the hooked ends, hang pieces of light, thin tin or aluminum foil. The foil should be trimmed to about three eighths of an inch wide and one inch long. Wrap the bent copper wire with plastic tape for insulation. Now you are ready to try it out.

Two convenient materials can be used to produce an electrostatic charge:
1. A hard rubber or vulcanite rod rubbed with a wool cloth will charge the rod negatively. Electrons will be transferred from the wool to the rod by the rubbing action.
2. A glass rod rubbed by a piece of silk cloth will charge the glass rod positively. Electrons will leave the glass and move to the silk by the rubbing action.

If vulcanite and glass rods are not on hand, borrow them from the science department. You might also experiment with other kinds of material. Try your plastic fountain pen.

EXPERIENCE 1. Detect an electrostatic charge. Rub the vulcanite rod vigorously with the wool cloth to build up a negative charge. Bring the rod close to, BUT NOT TOUCHING, the electroscope rod. Note that the foil leaves in the jar move outward and try to get away from each other. See Fig. 2-6.

WHY? Existing around the negatively charged vulcanite rod is an invisible electrostatic field or force. When brought close to the top end of the electroscope, the electrons at the top are driven downward to the foil leaves. Both leaves become negative and repel each other, therefore spreading apart. LIKE CHARGES REPEL. Remove the vulcanite rod and note that the leaves return to their normal position.

EXPERIENCE 2. Charge the electroscope inductively. A new word, INDUCTIVELY

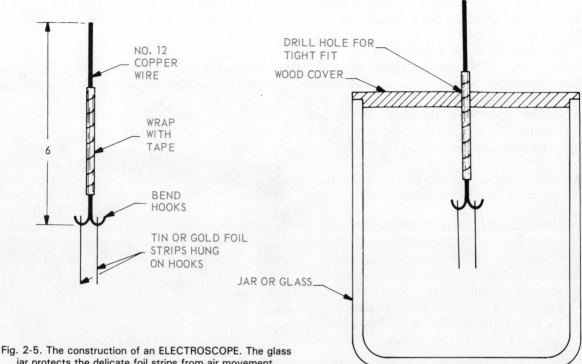

Fig. 2-5. The construction of an ELECTROSCOPE. The glass jar protects the delicate foil strips from air movement.

means: "by the influence of, or to cause it to become." Once again, charge the vulcanite rod and bring it close to the electroscope. The leaves will expand. Now, while holding the rod in this position, touch the top of the electroscope with your finger. The leaves will close. REMOVE

YOUR FINGER, THEN THE ROD. The leaves expand again and remain in this position. The electroscope is charged POSITIVELY.

WHY? The first action is the same as in the first experiment. However, when you touched the electroscope, electrons repelled by influence of the negative rod were also forced into your body through your finger. You removed your finger and when the rod was also removed the electroscope had no means of regaining the electrons which were forced into your body. Therefore the electroscope is left with fewer electrons. It is electrically out-of-balance and is positive. Both leaves are now positive and repel each other. LIKE CHARGES REPEL.

To further prove the Law of Electrostatic Charges, touch the electroscope with your finger. The leaves close and remain closed. The ELECTRONS from your body were returned to the electroscope and made it neutral.

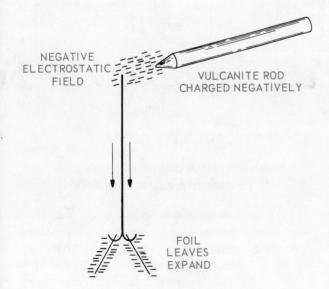

Fig. 2-6. Electrostatic field around charged rod drives electrons on electroscope rod to the foil leaves. Both leaves become negative and REPEL EACH OTHER.

EXPERIENCE 3. Charge the electroscope by contact. Only a weak charge on the vulcanite rod is required, so rub it lightly. Touch the electroscope with the charged rod, then remove it.

The leaves expand and remain expanded. The electroscope is charged NEGATIVELY.

WHY? Since the rod is negative with an excess of electrons and the electroscope is neutral, the total charge on the rod shares its electrons with the electroscope when contact is made. Of course, when the rod is removed, the electroscope cannot rid itself of these electrons and therefore remains charged NEGATIVELY.

EXPERIENCE 4. Repeat EXPERIENCES 1, 2 and 3 using the glass rod and silk.

EXPERIMENTER BOARD PROJECT

The Exploring Electronics Experimenter Board shown in Fig. 2-7 can be used by students to breadboard circuits before final construction. This useful project uses a five volt regulated power supply and a plug-in circuit board mounted on a plastic case. The ex-

perimenter can quickly connect circuits to see if they work properly without soldering components together.

Refer to Fig. 2-8 for the power supply and experimenter schematic board and parts list. In

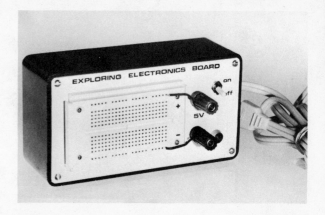

Fig. 2-7. Exploring Electronics Experimenter Board.

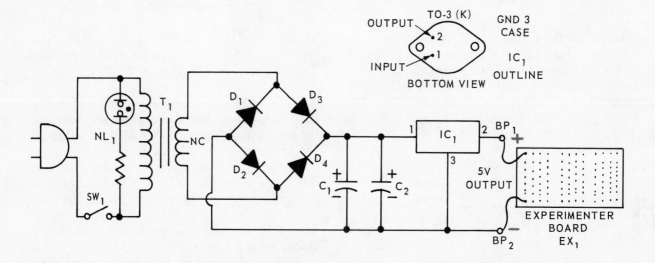

PARTS LIST FOR EXPLORING ELECTRONICS EXPERIMENTER BOARD

SW_1 — SPST switch

BP_1, BP_2 — 5-way binding posts

NE_1 — Neon lamp assembly or a neon lamp such as NE-51, a 100 KΩ, 1/2W resistor and a socket

T_1 — transformer, 117V ac primary, 6.3V secondary @ 1 amp

C_1, C_2 — electrolytic capacitors, 2200 μF @ 16 WVdc

D_1-D_4 — 1N 4001 diodes or equivalent

IC_1 — voltage regulator, LM 309K @ 5V

EX_1 — experimenter board (CSC, Experimenter Model 350)

Misc. — plastic case (3 1/8 × 6 × 2 3/16 in.) with cover, line cord with plug, screws, hookup wire, solder, decals

Fig. 2-8. Schematic and parts list for power supply and Experimenter Board.

Fig. 2-7, note the binding posts used to connect the power supply output to the experimenter board.

FORWARD STEPS TO UNDERSTANDING ELECTRICITY-ELECTRONICS

1. Electricity is a form of ENERGY.
2. Potential energy is energy at rest and has the capacity to do work.
3. Gravity is a universal law of nature.
4. Potential energy is energy stored as a result of position.
5. Matter is anything which occupies space.
6. All matter is made up of the basic elements of nature.
7. A molecule is a unit of matter.
8. Molecules are made of small particles called atoms. Each atom has its own characteristics.
9. A PROTON is a positively charged particle.
10. An ELECTRON is a negatively charged particle.
11. The nucleus of the atom usually contains protons and neutrons.
12. An electrically unbalanced atom becomes an ION.
13. Like charges repel each other.
14. Unlike charges attract each other.
15. Any charged object is surrounded by an invisible electrostatic field of force.

RULES OF SAFETY

SHOCKS ARE DANGEROUS

Electricity is a "shocking affair." You can get shocked when working around electrical equipment. Getting shocked is usually due to CARELESSNESS. Always THINK BEFORE ACTING.

HAND TO HAND INCREASES DANGER

In electricity, if you take hold of a live circuit with both hands, current flows across your chest and heart. This is not a safe practice. The TV technician, working with high voltages, generally keeps one hand away from the work to be tested. It may be inconvenient, but it is safer to work with one hand only.

TEST YOUR KNOWLEDGE - UNIT 2

Write your answers on a separate sheet of paper. Do not write in this book.

1. Write out in your own words the meaning of:
   Electricity
   Potential Energy
   Electron
   Negative Ion
   Inductively Charged
2. An electroscope has been charged negatively. What will happen if a positive rod is brought close to it?
3. How can the electroscope be used to discover if a body is charged negatively or positively?
4. In building the electroscope, why is the center wire wrapped with tape?
5. In Fig. 2-9, two pith balls are suspended on threads. What will happen if both balls are charged by contact with a charged vulcanite rod?
6. In Fig. 2-9, what will happen if ball A is charged negatively and ball B positively? Explain.
7. In studying the structure of an atom, if the number of protons is equal to the number of electrons, they are said to be equal in number and opposite in_____.
   Therefore, the atom has no electrical charge and is considered to be_____.

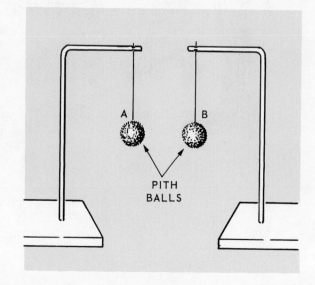

Fig. 2-9. Pith balls hung on strings to demonstrate static electrical charges.

This 40-foot long control panel is designed to monitor the
charging and polymerization of polyvinyl chloride (PCV).
(Firestone Tire & Rubber Co.)

Unit 3

THE LANGUAGE OF ELECTRICITY

The terms used in electronics are somewhat different from those we use in everyday language. This unit will help to explain important words, terms and instruments used in the field of electronics.

Major topics covered are:
1. What voltage is, and how it can be measured.
2. How current is produced by the movement of electrons.
3. How current can be measured.
4. What electrical resistance is, and the factors that affect it.
5. How to measure resistance, and how to read the value of resistors.
6. How to use the multimeter.

VOLTS AND VOLTAGE

In Unit 2, you read about potential energy. This was defined as "energy at rest." Potential energy, you learned, has the capacity to do work. You also performed experiments to prove that atoms can become negative or positive IONS by the transfer of electrons from generous to greedy atoms.

Now look at the two balls or terminals in Fig. 3-1. One has been charged in a negative direction; the other has been charged in a positive direction. One has many MORE electrons than protons; the other has many LESS electrons than protons. We say that the body of the terminal on the LEFT (A) is NEGATIVE. It is marked with a minus sign (−). The terminal on the RIGHT (B) is POSITIVE and is marked with a positive sign (+).

A source of potential energy is created when unequal quantities of electrons are stored at each terminal. In mechanics, this potential energy is measured in foot-pounds. In electricity, the potential is measured in VOLTS. This term was so named to honor Alessandro Volta, an Italian scientist.

The letter symbol for volt is E or V. When electricians speak of VOLTS or VOLTAGE, they talk about a POTENTIAL DIFFERENCE

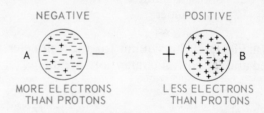

Fig. 3-1. Charged bodies or terminals have a potential difference. Note that terminals have unequal quantities of electrons, creating a source of potential energy.

between two terminals. It may also be referred to as a FORCE or an ELECTROMOTIVE FORCE (EMF). Because this voltage or potential difference or EMF exists, the source of energy has the capacity to do work.

The term VOLT probably is not new to you. At home, the lights and appliances operate on 117 volts. The family car has a 12 volt battery. Your transistor radio may use a 9 volt battery. Your flashlight may use two or three 1.5 volt cells. The exact value of a volt, as we measure it, has been established by international standards. It is the same anywhere in the world.

IMPORTANT: VOLTAGE IS ALWAYS MEASURED BETWEEN TWO POINTS IN AN ELECTRICAL CIRCUIT. THE VOLTAGE IS THE DIFFERENCE BETWEEN THE TWO POINTS.

Voltage is static (nonmoving) pressure only. At home, the water pressure at the faucet in the kitchen sink may be 20 lb. per square inch. NO WATER IS FLOWING. Yet, the pressure is always available to force out water if you turn on the faucet.

HOW TO MEASURE VOLTAGE

The instrument used to measure voltage is called a VOLTMETER. It has two test leads. The test leads usually are red and black. The RED lead is the POSITIVE LEAD and is connected to the positive terminal (highest potential point in the circuit). The BLACK lead is the NEGATIVE LEAD and must be connected to the negative terminal (lowest potential point in the circuit). A voltmeter must be connected across (in parallel) the voltage to be measured.

NOTE: Failure to connect a voltmeter correctly may destroy the meter.

In Fig. 3-2, the symbol for the voltmeter is used and the correct connections are made.

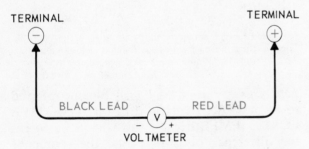

Fig. 3-2. A voltmeter must always be connected with black negative lead to negative and red positive lead to positive.

The value of the voltage or potential difference measured is read on the voltmeter scale. A simple voltmeter scale is illustrated in Fig. 3-3. Note that the scale reads from zero to 15 volts. At one third of the scale, there is a 5 volt mark. At two thirds of the scale, there is a 10 volt mark.

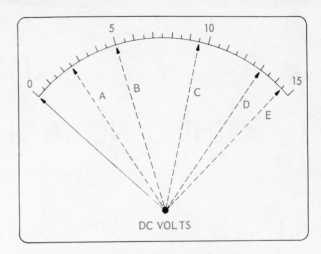

Fig. 3-3. Voltmeter scale for 15 volts. Pointer in Position A = 2.5V; Position B = 5V; Position C = 9.5V; Position D = 13V; Position E = 14.5V.

Note that in between 0 and 5 and 10 and 15 on the scale are shorter indicating marks which have no numbers. Each of the longer marks has a value of ONE VOLT. Between each of the one volt marks are shorter marks. Each of these marks has a value of .5 or one-half volt. With this meter scale arrangement, you may measure any voltage quite accurately between zero and 15 volts in steps of .5 volts. Study each position of the indicating pointer in the illustration.

NOTE: A meter such as this example can ONLY measure up to 15 volts. If you attempt to measure a voltage higher than 15 volts, it may destroy the meter.

WRONG CONNECTIONS DESTROY METERS

A voltmeter must be connected with the right polarity. Black or negative lead to the negative or ground point in a circuit. Red or positive lead to the positive or high voltage point in the circuit. Failure to do so causes the needle to move in the wrong direction.

MULTIRANGE VOLTMETER

Quite often you will need to measure voltages much higher than 15 volts. For this purpose, a multirange voltmeter is used. It has two or more meter scales and a RANGE SWITCH. A typical scale, Fig. 3-4, has a single scale with two sets of values. The values to use are determined by the setting of the range switch. The range, of course, will be determined by how high a voltage you want to measure.

Again, in Fig. 3-4, look at the scale and each set of values. Scale 0 to 25 has major divisions marked at 5, 10, 15, 20 and 25. Then, each major division is further divided into ten parts. Each mark has a value of .5.

Scale 0 to 5 has major divisions marked at 1, 2, 3, 4 and 5. Then, each major division is divided into ten smaller parts. Each mark has a value of .1.

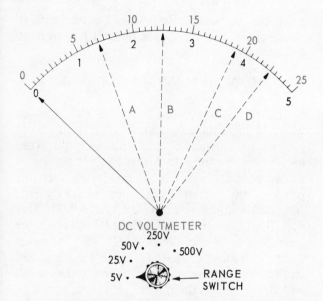

Fig. 3-4. The scale of a MULTIRANGE VOLTMETER with range switch. Needle is shown in positions A, B, C and D.

Which scale value should you use? Use the scale value divided to match the RANGE SWITCH POSITION. Study Fig. 3-5.

RANGE SWITCH POSITION	SCALE TO READ
5V	0-5
25V	0-25
50V	0-5
250V	0-25
500V	0-5

Fig. 3-5. Scale to read for range switch position.

Get the idea? It requires a little thought. To provide some practice readings, the indicating needle in Fig. 3-4 is shown in positions A, B, C and D. The values of the indicated voltage for

each range switch position appear in Fig. 3-6. It is suggested that you read the meter first, then compare your readings with the answers given in the table.

RANGE SWITCH POSITION		VOLTAGE VALUE AT INDICATOR POSITION (Refer to Fig. 3-4.)			
	Scale	A	B	C	D
5V	0—5	1.4	2.5	3.8	4.5
25V	0—25	7	12.5	19	22.5
50V	0—5	14	25	38	45
250V	0—25	70	125	190	225
500V	0—5	140	250	380	450

Fig. 3-6. Answers to voltmeter reading experiences.

USING PREFIXES

In our studies of electricity, we add a prefix or letters at the beginning of a word to change its meaning. Such is the case with VOLTS. See Fig. 3-7 for a table of these prefixes and their values. Frequently, the Greek letter μ (mu) is used for "micro." These prefixes are also used with other units of measurement.

1 megavolt (MV)	= 1 million (1,000,000) volts
1 kilovolt (KV)	= 1 thousand (1,000) volts
1 volt (V)	= 1 volt
1 millivolt (mV)	= 1/1,000 (.001) volt
1 microvolt (μV)	= 1/1,000,000 (.000001) volt

Fig. 3-7. Prefixes used in measuring voltage.

Another way of showing prefixes used for electronic units is shown in Fig. 3-8.

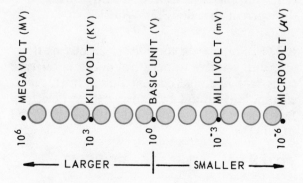

Fig. 3-8. Conversion chart for VOLT prefixes.

USE CORRECT SCALES FOR METER SAFETY

If you need to measure 100 volts, be sure to turn the "range selection switch" to its proper position. When measuring UNKNOWN VOLTAGES, always start with the highest range. You can switch down to a lower voltage range if necessary.

AMPS AND AMPERAGE

Thus far, the electrons which created a potential difference or voltage have been kept in prison, so to speak. They have had no freedom to run about. Yet, we did state that a potential voltage has the capacity to do work.

The two charged bodies or terminals shown in Fig. 3-1 have been redrawn in Fig. 3-9 with one very significant change. A copper wire has been connected between them. This copper wire is a CONDUCTING PATH for the electrons which will move from negative to positive until the charge on ball A is equal to the charge on ball B and NO POTENTIAL DIFFERENCE exists.

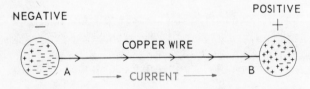

Fig. 3-9. The difference in potential causes a current to flow when a conducting wire is connected between the terminals.

This movement of electrons over a conducting path is called CURRENT. Its letter symbol is I. Current is measured in AMPERES. It is this current that does the WORK.

NOTE: Some texts consider electricity as flowing from positive to negative, and it is referred to as "conventional current flow." In this text, current will always be considered as "electron flow" and will move from NEGATIVE TO POSITIVE.

POTENTIAL DIFFERENCE REQUIRED

Again referring to the mountain lake in Unit 2, what would happen if the lake had no way to replace the water? Soon the lake would be dry. The potential energy provided by the water stored in the lake would be gone. The same condition results in Fig. 3-9. Current flows as long as a potential difference exists. If it is not RECHARGED, we have a DEAD SOURCE OF ENERGY. No current. No work can be done. THERE MUST BE A POTENTIAL DIFFERENCE FOR CURRENT TO FLOW.

SAFETY IN USE OF METERS

Meters are "the seeing eyes" of the technician. They are delicate and precise instruments. This means careful handling. Do not slam the meter down on the bench. Be gentle. Do not drop it on the floor. Meters should be kept free of dust and dirt.

HOW TO MEASURE CURRENT

The Forestry Service measures the flow of water in mountain streams by determining the gallons of water per minute which pass the measuring station. In electricity, the flow is measured by a specified number of electrons passing any given point per second. This specified amount is called a COULOMB. This specified amount or COULOMB flowing per second is equal to ONE AMPERE. A meter used to measure CURRENT must be connected so that ALL of the current flowing in the circuit will flow through the meter. This is expressed as "a METER CONNECTED IN SERIES in the circuit."

Fig. 3-10 shows an AMMETER connected in series. Note the symbol used to represent this meter.

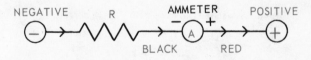

Fig. 3-10. The AMMETER or CURRENT METER must be connected in series in the circuit.

PREFIXES USED

Current prefixes are the same as those used with voltage. However, the terms kilo and mega are seldom used with amperes. Most currents which you will measure will be in milliamperes (one thousandth of an ampere) and amperes. Study the table in Fig. 3-11.

The conversion chart for the most common current prefixes is shown in Fig. 3-12.

1 ampere (AMP)	= 1 ampere
1 milliampere (mA)	= 1/1,000 (.001) ampere
1 microampere (μA)	= 1/1,000,000 (.000001) ampere

Fig. 3-11. Prefixes used in measuring current.

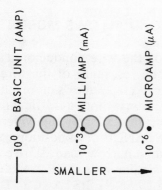

Fig. 3-12. Conversion chart for AMP prefixes.

HOW TO READ AN AMMETER

Like the voltmeter, the AMMETER has two test leads. The BLACK or NEGATIVE lead is connected to the negative side of the circuit. The RED or POSITIVE lead is connected to the positive side of the circuit.

NOTE: In order to measure current, the circuit must be disconnected and the meter connected in series

If connected wrong, the meter may be destroyed.

Now look at the simple milliammeter scale shown in Fig. 3-13. The scale reads 0, 50, 100 and 150. The greatest current you can measure with this scale is 150 milliamperes. The unnumbered marks between the major scale divisions represent 25, 75 and 125 milliamperes. These divisions are further divided by smaller marks representing 5 mA (note abbreviation) each. The pointer in Fig. 3-13 is shown in four positions. Study the readings.

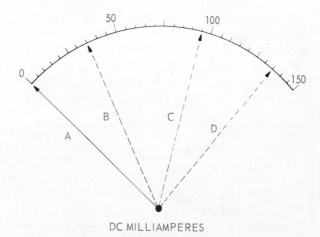

Fig. 3-13. Scale for simple MILLIAMMETER. Pointer in Position A = 0 mA; Position B = 35 mA; Position C = 95 mA; Position D = 135 mA.

AMMETER HOOKUP

To measure current flowing in a circuit, the conductor must be disconnected and the ammeter reconnected in SERIES with the circuit to be measured. The ammeter has very low internal resistance and should not disturb the circuit resistance.

MULTIMETERS

Frequently, currents smaller and larger than 150 mA must be measured. For these you will need a multirange ammeter. In Fig. 3-14, the

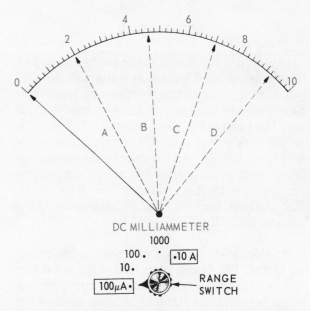

Fig. 3-14. MULTIRANGE AMMETER scale and range switch.

scale of a typical meter is sketched. It is easy to read since the scale reads only 0 to 10. The value of the scale will be determined by the RANGE SWITCH. Ranges on this meter are 100 μA (microamperes), 10 mA, 100 mA, 1000 mA and 10 A. Study the table, Fig. 3-15.

RANGE SWITCH POSITION	SCALE READS	EACH SMALL MARK EQUALS
100 μA	0 to 100 μA	2 μA
10 mA	0 to 10 mA	.2 mA
100 mA	0 to 100 mA	2 mA
1000 mA	0 to 1000 mA	20 mA
10 A	0 to 10 A	.2 A

Fig. 3-15. Range switch position and scale value.

The indicating needle is shown in four positions in Fig. 3-15. Read the meter and compare your results to the answers given in the table in Fig. 3-16.

RANGE SWITCH POSITION	CURRENT VALUE AT INDICATOR POSITION			
	A	B	C	D
100 μA	20 μA	46 μA	70 μA	92 μA
10 mA	2 mA	4.6 mA	7 mA	9.2 mA
100 mA	20 mA	46 mA	70 mA	92 mA
1000 mA	200 mA	460 mA	700 mA	920 mA
10 A	2 A	4.6 A	7 A	9.2 A

Fig. 3-16. Answers to ammeter reading experiences.

RESISTANCE

If you turn the faucet in the kitchen sink just a little bit, a small stream of water will flow. If you turn it more, a larger stream of water will flow. We can express this another way. As you turn the faucet ON, there is less RESISTANCE to the flow of water and therefore more water will flow. RESISTANCE may be defined as an "opposing or retarding force." In electricity, resistance is the opposition to the flow of an electrical current by a material, substance, component or device.

In electricity, the unit of resistance is called an OHM. The Greek letter omega (Ω) is used as the symbol for ohms. Example: one hundred ohms is written as 100 Ω. The letter symbol for resistance is R.

MATERIALS THAT CONDUCT ELECTRICITY

Materials have various abilities to conduct electricity. This is due, in part, to their atomic structure and the availability of "free electrons" in the material for conduction purposes. Silver is one of our best conductors. Copper is a good conductor and cheaper than silver. Iron and aluminum are good, too, but not as good as copper. Materials such as glass and rubber, which are not good conductors, are called INSULATORS.

A WIDER ROAD ALLOWS MORE TRAFFIC

The size of the wire conductor used determines the total resistance of the wire. As you know, a wide highway allows more automobiles to travel per hour without a traffic jam. We can say, therefore, that a larger and wider road has LESS RESISTANCE or opposition to the flow of traffic than a narrow road. Wire conductors behave in a similar way. THE LARGER THE WIRE, THE LESS RESISTANCE IT HAS TO THE FLOW OF ELECTRONS.

LONG ROAD HAS MORE RESISTANCE

If a particular section of narrow road opposes the flow of traffic a certain amount, a section of narrow road four times that long could be expected to oppose the traffic four times as much. With electrical conductors, it is the same story. If a wire 10 ft. long has 1 ohm of resistance, then 100 ft. of the same wire would have ten times as much resistance, or 10 ohms. Longer wires must also be larger wires if they are to carry the required amount of CURRENT.

HEAT

Most materials used as conductors will increase in resistance if allowed to become warm. If wires are to be enclosed in conduit or pipe where they cannot radiate the heat into space, larger wires of less resistance should be used.

FACTS ABOUT RESISTANCE

Basically, we have learned that:

1. Resistance is measured in ohms.
2. The Greek symbol for ohms is Ω.

3. The letter symbol for resistance is R.
4. Resistance depends upon the material used to make the conductor.
5. Larger conductors have less resistance.
6. Longer wires have more resistance.
7. Heat will increase the resistance of a wire.

HOW TO MEASURE RESISTANCE

Resistance is measured with an OHM-METER. Whenever resistance is placed in a circuit, it is represented by this symbol:

With this in mind, see Fig. 3-17:

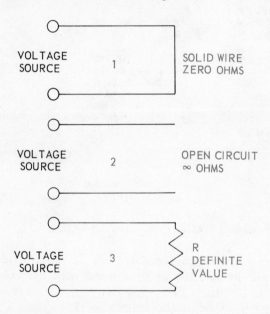

Fig. 3-17. Three conditions can exist when measuring resistance in a circuit.

1. A wire is connected directly across a voltage source. The wire is so short, it is assumed to have ZERO RESISTANCE. This is called a SHORT CIRCUIT, and maximum current will flow.

2. Here, two wires are connected to a voltage source, but the wires are not connected together. This is called an OPEN CIRCUIT. There is NO current path and no current will flow. Its resistance is infinity. The symbol for infinity is ∞. Using symbols, the circuit has R = ∞ ohms.

3. In this circuit, a definite value of resistance, represented by the symbol R, has been placed in wires between the voltage terminals. In this case, the resistance is not zero, not infinity, but of some value. Current will flow in the circuit, depending on the value of the voltage and resistance.

OHMMETER SCALE

The ends of a scale used to measure resistance must be at zero and at infinity. All points between these two limits will have some definite value. Fig. 3-18 shows a typical scale used on the OHMMETER, RANGE SWITCH and a knob called OHMS ADJUST.

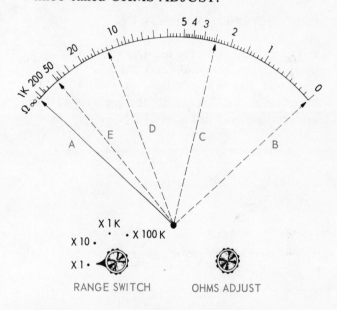

Fig. 3-18. Typical OHMMETER SCALE with range switch and ohms adjust knob.

Before measuring ohms, the meter must be ADJUSTED TO ZERO. Proceed as follows: Take the two test leads and touch them together. This is ZERO resistance. Adjust meter to zero with ohms adjust knob.

Ohms will be read on the scale directly, but must be multiplied by the setting of the range switch. See Fig. 3-19.

SCALE × 1	value as read
SCALE × 10	multiply reading by 10
SCALE × 1K	multiply reading by 1000
SCALE × 100K	multiply reading by 100,000

Fig. 3-19. Multiply OHMMETER SCALE reading by setting of the range switch.

RANGE SWITCH POSITION	VALUE IN OHMS AT INDICATOR POSITION (Refer to Fig. 3-18.)				
	A	**B**	**C**	**D**	**E**
× 1	open circuit	short circuit	2.5 Ω	12 Ω	50 Ω
× 10			25 Ω	120 Ω	500 Ω
× 1K	∞	zero	2500 Ω	12 KΩ	50 KΩ
× 100K	ohms	ohms	250 KΩ	1.2 MΩ	5 MΩ

NOTE: K means KILO or 1000. M means MEGA or 1,000,000.

Fig. 3-20. Answers to ohmmeter reading experiences.

In Fig. 3-18, the needle is shown in five different positions. Read the meter and compare your answers with the table in Fig. 3-20.

PREFIXES USED IN RESISTANCE

The prefixes used for resistance are similar to the ones used for voltage and current. However, the most common prefixes for resistance are kilo and mega. See resistance prefixes in the conversion chart in Fig. 3-21.

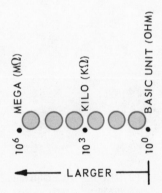

Fig. 3-21. Conversion chart for resistance prefixes.

OHMMETERS ARE SELF-POWERED

An OHMMETER has its own dry cells inside the meter case. Therefore, it is self-powered and should not be connected to a piece of equipment that is "turned on." When measuring resistances in a piece of equipment, turn off the power and/or pull out the plug from the power source. Accidentally connecting an ohmmeter to a "live" circuit could damage the meter.

The dry cells in an ohmmeter will grow weak with age or continued use. This will be revealed if you cannot adjust the meter to zero.

Therefore, before measuring resistance, always bring the test leads together (zero resistance) and adjust zero on the meter scale.

RESISTOR COLOR CODE

In your electronics studies, you will be using various kinds of resistors. Many carbon type resistors have bright colored bands. These bands will tell you the value of the resistors in ohms. The resistor COLOR CODE, Fig. 3-22, is presented for your study. This should be memorized.

COLOR	NUMERICAL FIGURE	MULTIPLIER	TOLERANCE
BLACK	0	× 1	
BROWN	1	× 10	
RED	2	× 100	
ORANGE	3	× 1000	
YELLOW	4	× 10,000	
GREEN	5	× 100,000	
BLUE	6	× 1 Million	
VIOLET	7	× 10 Million	
GRAY	8	× 100 Million	
WHITE	9	× 1000 Million	
SILVER	-	× .01	±10%
GOLD	-	× .1	±5%
NONE	-	- - -	±20%

Fig. 3-22. Resistor color code.

READING RESISTOR COLOR CODE

To identify a resistor from the color code, hold the resistor in your hand with the color bands on the LEFT. The first band color is the first number of the value. See Fig. 3-23. The second band color is the second number of the value. The third band color tells you to multiply the first two numbers by this factor.

In Fig. 3-23, a resistor with bands of BROWN, BLACK, GREEN and SILVER represents 1,000,000 Ω or 1 megohm.

The fourth band, silver, tells you how accurate the resistor must be in order to pass inspection. This resistor is ± 10 percent. It could actually measure as high as 1.1 megohms or as low as 900 KΩ and still be acceptable (10 percent above to 10 percent below specified value).

The more accurate a resistor is, the more expensive it becomes. Most expensive equipment will use ± 5 percent resistors and ± 1 percent resistors. Working with the color code is the best way to learn it. You will be given many chances to practice.

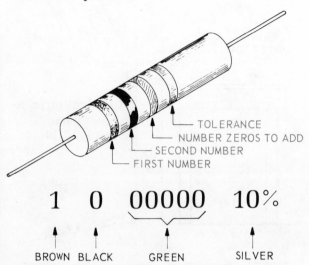

Fig. 3-24. A typical VOLT-OLM-MILLIAMMETER (VOM) is a combination meter to measure volts, ohms and amperes. (Triplett)

1	0	00000	10%
BROWN	BLACK	GREEN	SILVER

Labels on resistor: TOLERANCE / NUMBER ZEROS TO ADD / SECOND NUMBER / FIRST NUMBER

Fig. 3-23. How to read the resistor color code.

USING A MULTIMETER

For convenience, the three meters you have learned to read (voltmeter, ammeter, ohmmeter) are available in one case. It is called a VOLT-OHM-MILLIAMMETER (VOM). See Fig. 3-24. The center switch must be used to select WHAT YOU WANT TO MEASURE (volts, amperes or ohms) and WHAT RANGE YOU WISH TO USE.

DIGITAL MULTIMETERS

With advances in technology, digital multimeters are becoming more popular with students, hobbyists and service technicians. A digital multimeter is shown in Fig. 3-25.

RAIN DETECTOR PROJECT

Have you ever wondered whether or not it was raining, at night or when you were unable to look outside? If so, build this rain detector project, Fig. 3-26. It will detect rain drops or water, so it also can be used as a liquid level detector.

In wiring the circuit, it is a good idea to use a socket for the IC (integrated circuit). Fabricate

Fig. 3-25. A DIGITAL MULTIMETER simplifies electrical tests. (Heath Co.)

Fig. 3-26. Rain Detector.

the rain sensor printed circuit as shown in the schematic and parts list, Fig. 3-27. The sensor must be placed outside, Fig. 3-28, so the rain drops can fall on it.

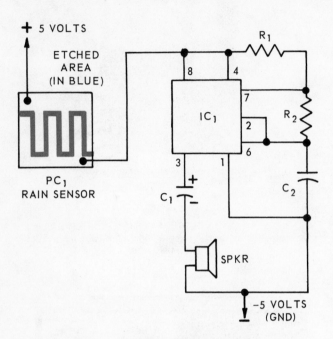

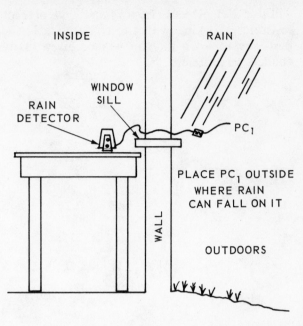

Fig. 3-28. Connections and installation of Rain Detector.

PARTS LIST FOR RAIN DETECTOR

R_1 — 4.7 KΩ, 1/2W resistor
R_2 — 150 KΩ, 1/2W resistor
C_1 — electrolytic capacitor, 4.7 μF @ 16 WVdc
C_2 — capacitor, .01 μF @ 100 WVdc
IC_1 — integrated circuit, LM 555 or NE 555P or RS 555
SPKR — 2 in., 8 Ω speaker
PC_1 — 1 in. square (etched as shown)
Misc. — power supply, 5V or 3-1 1/2V "C" cells, IC socket, wire, plastic case, PC board and materials, solder, decals

Fig. 3-27. Schematic and parts list for Rain Detector.

The effect of the rain drops on the printed circuit area is that the moisture provides a low resistance connection across the contacts. This connection is monitored by the IC, and it triggers an alarm signal through the speaker. A 5 volt power supply or three 1 1/2 volt "C" cells can be used to power the rain detector. The plastic case used to house this unit is made from an inexpensive watch case.

STEPS TO UNDERSTANDING ELECTRICITY-ELECTRONICS

1. A VOLT is the unit of measurement of electrical pressure. Its letter symbol is E or V in drawings or equations..
2. POTENTIAL DIFFERENCE (electromotive force) is measured in volts.
3. Voltage is always measured between two points in an electrical circuit.
4. The negative or black lead of a voltmeter must always be connected to the point of lowest or most negative potential.
5. The flow of electricity is called CURRENT. Its letter symbol is I. It is measured in amperes and milliamperes.
6. The ammeter must always be connected in SERIES in the circuits so that all current will flow through it.
7. The resistance to the flow of electricity is measured in OHMS. Its letter symbol is R. The Greek letter omega (Ω) means OHMS.
8. A short circuit represents ZERO OHMS.
9. An open circuit reads infinity (∞) ohms.
10. Before measuring ohms, the ohmmeter must be adjusted to zero.
11. An ohmmeter is self-powered. Never connect its test leads to a "live" circuit.
12. A multimeter measures volts, amperes and ohms in several ranges.

TEST YOUR KNOWLEDGE - UNIT 3

1. Select five resistors and record this information:

	COLOR CODE	VALUE IN OHMS	TOLERANCE
R_1	_____	_____ Ω	_____
R_2	_____	_____ Ω	_____
R_3	_____	_____ Ω	_____
R_4	_____	_____ Ω	_____
R_5	_____	_____ Ω	_____

2. Ask your teacher for an ohmmeter. Measure each of your resistors and record their values. Are they within their range of tolerance?

	MEASURED VALUE	ACCEPTABLE WITHIN TOLERANCE
R_1	_____	Yes or No _____
R_2	_____	Yes or No _____
R_3	_____	Yes or No _____
R_4	_____	Yes or No _____
R_5	_____	Yes or No _____

Did you ZERO the ohmmeter before making your ohms measurements?

3. LEARNING TO USE A VOLTMETER. Refer to Fig. 3-29. Connect the voltmeter to a variable voltage supply (0-15 volts). Set voltage control knob to minimum and TURN ON the supply. Adjust control knob to obtain readings of 1 volt, 3.5 volts, 5 volts, 8 volts, 11 volts, 13.5 volts and 15 volts on the voltmeter.

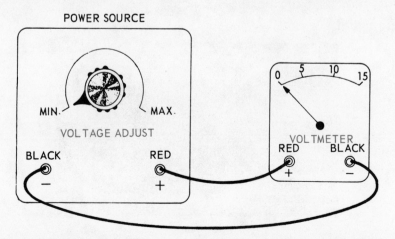

Fig. 3-29. Power supply and voltmeter connections for learning to use a voltmeter.

4. Disconnect your voltmeter from the power supply. Turn the control knob to any position. Now, reconnect the voltmeter and record the voltage. Disconnect voltmeter again and reset control knob at another position. Measure the voltage. Repeat this experience several times until you can measure and read any voltage.

5. What is your body resistance? Hold one probe of the ohmmeter between the fingers of your right hand and the other probe in your left hand. What is your resistance to the flow of current? Now, wet your fingers and repeat the measurement. Is there a difference? Is it safe to work around electricity in wet places?

6. LEARNING TO USE AN AMMETER. In this experience you will need a LOAD that will draw some current from this power source. Select a 100 ohm resistor and connect it in a circuit as shown in Fig. 3-30. Start from the negative terminal and trace the path of electron flow through the RESISTOR, through the AMMETER and then to the positive terminal of the power source. Turn on the power souce and increase its voltage to maximum value. Note the CURRENT value on the milliammeter.

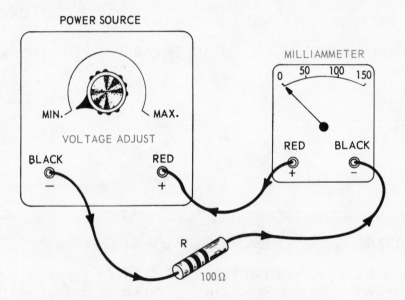

Fig. 3-30. Connection of circuit to measure CURRENT.

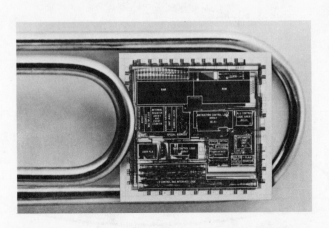

A one-chip computer is compared to a standard paper clip.   (Bell Laboratories)

Unit 4

POWER AND WORK

In Unit 3, you learned about the VOLT, the AMPERE and the OHM. You found that voltage, current and resistance are key factors in electricity and electronics.

In this Unit, you will learn:

1. What relationship exists among voltage, current and resistance.
2. What electrical power is, and how it can be computed.
3. How to read a kilowatt-hour meter.

VOLTAGE, CURRENT AND RESISTANCE

A diagram of the circuit and meters to be used in the following problems are shown in Fig. 4-1. The schematic diagram appears in Fig. 4-2.

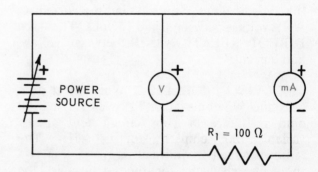

Fig. 4-2. Schematic diagram for the circuit in Fig. 4-1.

SOLVING PROBLEM 1. How does a change in VOLTS affect an electrical circuit? Turn on the power source and adjust the voltage to 5 volts. What is the current? Now increase the

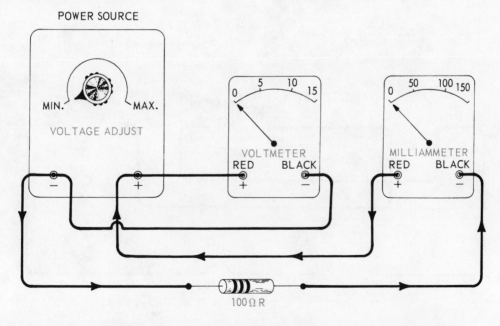

Fig. 4-1. Circuit set up to determine the relationship between VOLTS and AMPERES.

37

voltage to 10 volts. Read the milliammeter. Increase the voltage to 15 volts and read the milliammeter. What are your conclusions?

SOLUTION: As you might expect, as the pressure is increased, the flow is increased. IN ELECTRICITY, IF THE VOLTAGE IS INCREASED, THE CURRENT WILL INCREASE. ALSO, IF VOLTAGE IS DECREASED, THE CURRENT WILL DECREASE.

In Fig. 4-3, a GRAPH shows you how the engineer expresses this relationship. By using this type of graph, you can find the current at ANY voltage between 0 and 15 volts. There is a DIRECT RELATIONSHIP between voltage and current.

SOLVING PROBLEM 2. What effect does resistance have on a circuit? Leave the fixed 100 ohm resistance in your circuit and add a variable 1000 ohm resistor in series. The redrawn circuit is shown in Fig. 4-4. Fig. 4-5 shows a schematic for the circuit. Set the voltage at 15 volts and observe the current. Now, turn the knob of the variable resistor TO INCREASE ITS RESISTANCE. What hap-

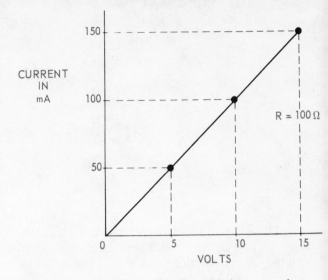

Fig. 4-3. A graph to show the relationship between voltage and current. Resistance is 100 ohms.

pens to the current? Turn the variable resistance in the opposite direction TO DECREASE ITS RESISTANCE. What happens to the current?

SOLUTION: Here is proof of another basic fact about electrical circuits. If you DECREASE THE RESISTANCE THERE IS INCREASED CURRENT FLOW. IF

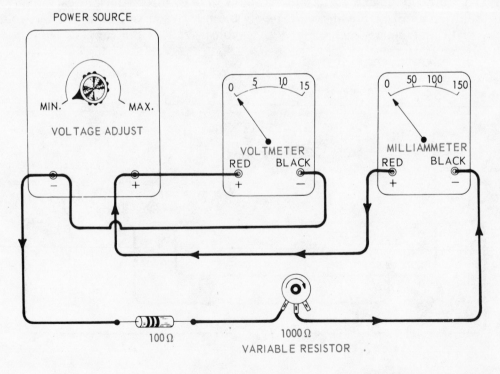

Fig. 4-4. Circuit which may be used to prove the relationship between current and resistance in a circuit.

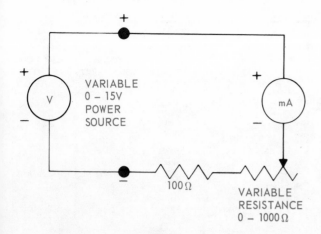

Fig. 4-5. The schematic for circuit Fig. 4-4.

RESISTANCE IS INCREASED, THE CURRENT MUST DECREASE.

This relationship may be expressed in an upside down fashion. CURRENT IS INVERSELY related to the RESISTANCE in a circuit:

R increases, I decreases

or

R decreases, I increases

These voltage, current and resistance relationships are expressed by the well known OHM'S LAW. This law states that:

$$I = \frac{E}{R} \text{ or } R = \frac{E}{I} \text{ or } E = I \times R$$

when I = amperes, R = ohms and E = volts.

You can see that if any two values are known, the third and unknown can be found by simple mathematics.

WORK

We learn about WORK at any early age. Mow the lawn. Run errands. Carry some wood. Study. All these jobs are work and require physical effort and the use of ENERGY. Lift a ten pound stone to a height of one foot. You have done ten foot-pounds of work. When a force moves through a distance, WORK is done. The definition in physics will say that:

$$\text{work} = \text{force} \times \text{distance}$$

Look at a lawn mower. Assume it requires 10 pounds of force to push it through the grass. If your lawn is fifty feet wide, then you would do 500 foot-pounds of work by pushing it once across the lawn. It takes a lot of work to mow a lawn. Energy is used up.

POWER

Power is defined as the RATE OF DOING WORK. How long does it take you to push the lawn mower across the lawn just once? One minute? Ten minutes? Thirty seconds? In the first example, you would do 500 ft. lb. of work per minute. In the second, 500 ft. lb. of work per 10 minutes and in the third example, 500 ft. lb. of work per 30 seconds.

James Watt, inventor of the steam engine, looked around for a way of determining the power of his newly invented engine. In those days, the farm horse did most of the work, so Watt compared his engine to the rate of work of the horse. As a result, we have the familiar term "horsepower." A one horsepower rating states that 33,000 foot-pounds of work are done in one minute or 550 foot-pounds per second.

WATTS

In electricity, we can now state that VOLTAGE is a force, and the movement of a quantity of electrons per second is an AMPERE. We arrive at the conclusion that VOLTS times AMPERES equals POWER. So that James Watt will not be forgotten, the unit of electrical power is named a WATT.

To find the rate of the use of energy or power of a circuit, use this formula:

VOLTS times AMPERES equals WATTS

or

$$E \times I = P \text{ in watts}$$

WATT-HOURS

We have been reading about the rate of using electrical energy, but another point must be considered. How long did you work at that rate? If you worked for one hour, you can determine the total energy used. If an electrical device such as a lamp or a motor is using power at the rate of 100 watts, in one hour it would have used 100 WATT-HOURS of electrical energy. The power company sells power to your home at an established rate per KILOWATT-HOUR (KWH). The meter measures the amount of power used. At the end of the month, a bill is sent.

EXAMPLE 1. Assume that electricity costs five cents per KWH in your locality. At night your family burns ten 100 watt lights for five hours. How much does it cost?

$$100 \text{ watts} \times 10 \times 5 = 5000 \text{ watt-hours}$$
$$5000 \text{ watt-hours} = 5 \text{ KWH}$$
$$5 \text{ KWH} \times 5 \text{ cents} = 25 \text{ cents}$$

EXAMPLE 2. Your home has a voltage of 117 volts. A toaster draws 5 amperes of current and you toast bread for 30 minutes. How much does it cost if energy costs 5 cents per KWH or .005 cents per watt-hour?

$$P = I \times E \text{ so}$$
$$WH = 585W \times .5 \text{ hrs.} = 292.5 \text{ watt-hours}$$
$$292.5 \text{ watt-hours} \times .005 \text{ cents} = 1.46 \text{ cents}$$

EXAMPLE 3. The water heater in your home has a 2000 watt heating element. At a voltage of 117 volts how much current does it use?

$$I \times E = P \qquad I \times 117V = 2000 \text{ watts}$$
$$I = 17.1 \text{ amps}$$

See the table in Fig. 4-6 for a list of common appliances used in the home and their consumption of power in watts.

Power companies continue to supply an almost unlimited quantity of electrical energy. Some is generated by hydroelectric plants that convert the energy of falling water to electric power. In many locations, steam produced by coal, gas or oil-fueled boilers turns the generators. The use of atomic energy may become our major source of power in the future. of power in the future.

READING YOUR ELECTRIC METER

The arrival of the "meter reader" from the power company is generally announced by the

ELECTRICAL APPLIANCE	AVERAGE WATTAGE	KILOWATT HOURS CONSUMED ANNUALLY	ESTIMATED AVERAGE COST FOR 2 MONTHS OPERATION
Food			
Coffee maker	894	106	31¢
Dishwasher	1,201	363	44¢
Freezer (frostless, 15 cu. ft.)	440	1,761	$3.42
Frying pan	1,196	186	60¢
Range	8,200	1,175	$3.66
Refrigerator (frostless, 12 cu. ft.)	321	1,217	$3.17
Toaster	1,146	39	13¢
Laundry			
Clothes dryer	4,856	993	$2.95
Iron (hand)	1,008	144	45¢
Washing machine	512	103	33¢
Comfort			
Air conditioner (room)	1,566	1,389	$4.43
Electric blanket	177	147	44¢
Dehumidifier	257	377	$1.43
Entertainment			
Radio	71	86	15¢
Radio-record player	109	109	33¢
Television (black and white)	237	362	$1.06
Television (color)	332	502	$1.37

Fig. 4-6. Power consumption by appliances. (Based on estimates by the Electric Energy Association.)

dogs in the neighborhood. Sometimes over-protective dogs make meter reading a hazardous occupation. Now, new equipment is being tried that automatically reads meters by electronic circuitry and telephones this information to the business office computer of the power company. Fig. 4-7 shows a line crew at work.

Your house meter probably has four dials with numbers on each reading 1 to 10. See Fig. 4-8. The first dial on the right reads in kilowatt-hours. The second dial reads in ten kilowatt-hours. The third dial reads in 100 kilowatt-hours. The fourth dial reads in 1000 kilowatt-hours. NOTE: When reading any of the four dials, always use the number which the pointer has just passed. If the pointer is between 6 and 7, use 6. If the pointer seems to be on a number, refer to the dial on its right to determine if the pointer is exactly on the number or a little before or after the number. The examples in Fig. 4-8 will help you to understand.

Fig. 4-7. Electric company line crew services the power transmission lines. (Carolina Power & Light Co.)

Assume that one month ago your meter read 3152. Today, it reads 4827. The power consumed for one month would be: 4827 − 3152 or 1675 KWH. At a typical cost of 5 cents per KWH, the electric bill would be 1675 × .05 = $83.75. However, the rate usually changes to a lower cost per kilowatt-hour if you use a large quantity of power. For example:

First 50 KWH at .05
Next 100 KWH at .03
All above 150 KWH at .02

Using these rates, the bill would be:

50 KWH at .05 = 2.50
100 KWH at .03 = 3.00
1525 KWH at .02 = 30.50
Total cost $36.00

CONTINUITY CHECKER PROJECT

The audible continuity tester, Figs. 4-9 and 4-10, can be used to test for "opens" and "shorts" in a circuit. The tester is built around a 555 integrated circuit, which is used as an audible sound source. Touch the test probes together. If the circuit is energized, a tone will be heard.

To test for shorts, turn off all power from the circuit to be tested. Connect the two leads to the circuit. If a tone is heard, there is a complete short. If the circuit is open, there will be no audible sound. This device also can be used to test switches, fuses and circuit breakers.

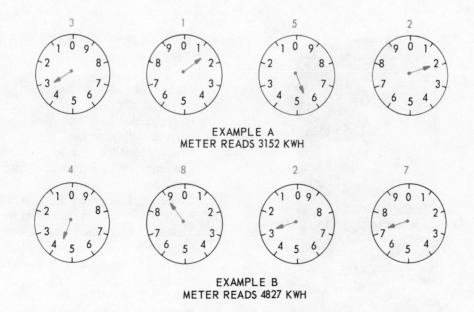

EXAMPLE A
METER READS 3152 KWH

EXAMPLE B
METER READS 4827 KWH

Fig. 4-8. Examples of meter reading: A Meter reads 3152 KWH. B Meter reads 4827. Note in Example B that dial 2 (second from right) reads 2, not 3. It will not be 3 until dial 1 reaches 0. The same with dial 3. It will not be 9 until dial 2 reaches 0.

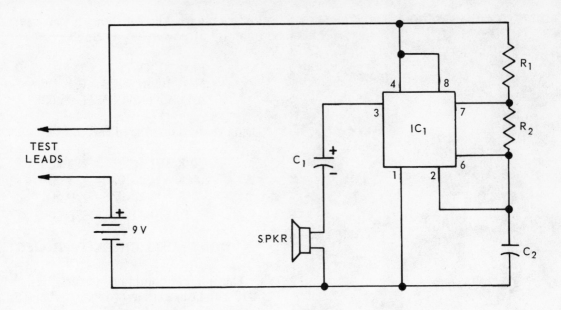

PARTS LIST FOR AUDIBLE CONTINUITY CHECKER

R_1 — 4.7 KΩ, 1/2W resistor
R_2 — 150 KΩ, 1/2W resistor
C_1 — electrolytic capacitor, 4.7 μF @ 16 WVdc
C_2 — capacitor, .01 μF @ 100 WVdc
IC_1 — integrated circuit, LM 555 or

NE 555P or RS 555 (or equivalent)
SPKR — 2 in., 8 Ω speaker
Misc. — plastic case with cover, alligator clips with insulated covers, wire, PC board and materials, solder, screws, IC socket, 9V battery

Fig. 4-9. Schematic and parts list for Audible Continuity Checker.

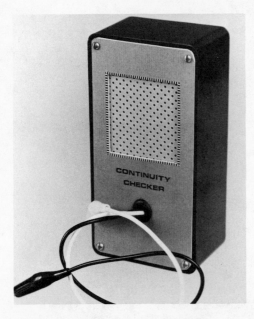

Fig. 4-10. Audible Continuity Checker.

FORWARD STEPS IN UNDERSTANDING ELECTRICITY-ELECTRONICS

1. As voltage of a circuit increases, current will also increase.
2. If resistance in a circuit is increased, the current will decrease.
3. Voltage and Current are directly related.
4. Current and Resistance are inversely related.
5. Power is the rate of doing work.
6. Electrical power is measured in watts.
7. Voltage multiplied by Current equals Power.
8. Energy consumed is measured in watt-hours.
9. A kilowatt-hour equals one thousand watt-hours.
10. Electricity is purchased by the KWH unit of measure.

TEST YOUR KNOWLEDGE - UNIT 4

1. Read your electric meter today and again in one week. Inquire from your power company the cost of electric power per KWH. Figure the cost of the power used for the week.
2. A circuit has an applied voltage of 100 volts and a circuit current of 2 amperes. What is its power?
3. If your color TV uses 500 watts of power, how much does it cost to watch a two hour program at 5 cents per KWH?
4. The resistance of an electric roaster is 20 ohms. With a home voltage of 117 volts, how much power does the roaster use?
5. For Christmas decoration of your home, ten strings of colored lamps are used. Each string uses 50 watts of power. How much current will be required? Use voltage of 117 volts.
6. Explain what will happen if a short circuit should occur in a home appliance.
7. Connect the circuit with a variable resistor as in Fig. 4-4. Adjust voltage to maximum of 15 volts. Adjust resistance for a 100 mA current.
   a. Turn voltage to zero.
   b. Record current.
   c. Repeat in steps for 1, 2, 5, 10 and 15 volts. Record current after each step.
   d. Draw a graph of your results.

Among expanding career opportunities are installation, maintenance and repair of electrical and electronic devices and systems.

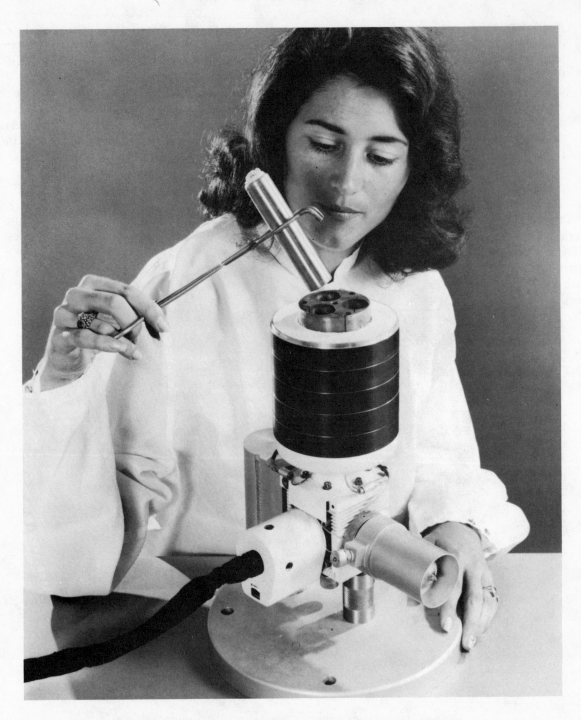

Nuclear energy produced by three capsules of plutonium-238 powers a 9 lb. refrigerator. This nuclear refrigerator will super cool microwave, laser and radiation detection systems as low as −320°F to increase their sensitivity. (U.S. Atomic Energy Commission)

Unit 5

SOURCES OF ELECTRICAL ENERGY

Through the ages, scientists have experimented with methods of creating a POTENTIAL DIFFERENCE that will cause electric current to flow. Engineers have applied their technical "know-how" to make the electric current perform useful work.

This Unit will explain:

1. How electricity is produced by chemical action.
2. What difference exists between certain types of cells and batteries.
3. How light energy can be changed to electrical energy.
4. How electricity can be produced from heat.
5. How electricity can be produced by pressure.

CHEMISTRY AND ELECTRICITY

The first practical electric BATTERY was demonstrated by the Italian scientist, Allessandro Volta, in about 1800. The battery consisted of pieces of zinc and silver separated by cardboard soaked in salt solution. See Fig. 5-1. Each pair of zinc and silver pieces is considered a CELL. Cells connected together as shown make a BATTERY.

The conclusion drawn from Volta's work is: If two dissimilar metals called ELECTRODES are placed in an electrolyte solution that chemically reacts with the metals, a POTENTIAL DIFFERENCE or VOLTAGE will be produced between the two metals. The chemical reaction causes one electrode to LOSE ELECTRONS and, therefore, becomes POSITIVE. The other electrode GAINS ELECTRONS and becomes NEGATIVE.

IN THE CELL, CHEMICAL ENERGY IS CONVERTED TO ELECTRICAL ENERGY.

EXPERIMENT 1. Secure a penny, a dime and a small piece of blotting paper. Soak the

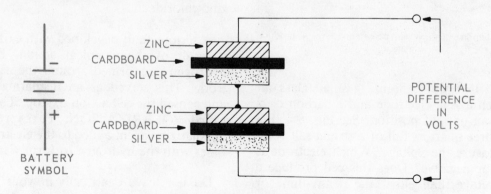

Fig. 5-1. The original BATTERY consisted of zinc and silver pieces of metal separated by cardboard soaked in salt solution.

45

blotting paper in salt water. Clamp the parts together with a clothespin as in Fig. 5-2. Measure the voltage with your meter. Is this an efficient cell? Is the dime or the penny negative?

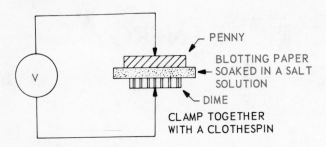

Fig. 5-2. A simple cell made with a penny and a dime.

EXPERIMENT 2. A lemon CELL. Secure a lemon or a grapefruit, a strip of zinc and a strip of copper. Stick the metal electrodes into the lemon as shown in Fig. 5-3. What voltage is developed? Which metal is negative? Which is positive? Try inserting the metal strips closer together. Does the voltage increase or decrease?

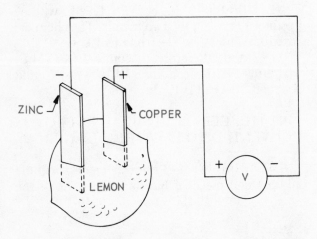

Fig. 5-3. A lemon cell made with copper and zinc will produce a voltage.

EXPERIMENT 3. Secure a small glass jar and fasten a zinc electrode and a carbon electrode in an upright position. See Fig. 5-4. Fill the jar three-quarters full of a strong salt solution. Measure the voltage. Which electrode is positive or negative? Does this cell produce a larger voltage than either the penny-dime or lemon cell? While the voltmeter is connected to the cell, connect a 100 ohm resistor to the cell.

The voltage of the cell will drop. With a load attached and drawing current, this inefficient cell cannot hold its voltage.

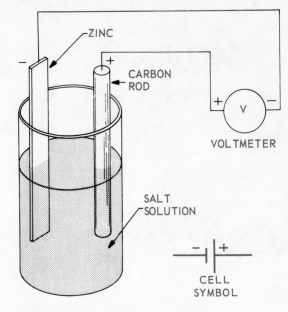

Fig. 5-4. A zinc-carbon cell will produce a voltage by chemical action.

ZINC CARBON CELLS

The flashlight CELL and the so-called DRY CELL are examples of cells which you may purchase from a hardware or electronics store. Typical cells are pictured in Fig. 5-5. You may compare the commercial cell with your experimental cell. Study the cutaway view in Fig. 5-6. Note that it is assembled in a ZINC can (zinc electrode). In the center is the carbon rod electrode. The electrolyte is made of absorbent material with a mixture of sal ammoniac and zinc chloride.

A serious fault developed with early cells. As a result of the chemical action, bubbles of hydrogen gas formed around the carbon electrode. This served as an insulating "blanket" and caused the cell action to stop. This defect is called POLARIZATION. To overcome it, a depolarizing agent added to the electrolyte combines with the hydrogen to form water.

Do not leave dead cells in your flashlight. Dead cells may expand and corrode as a result of chemical action and formation of water.

Fig. 5-5. Alkaline, mercury and silver-oxide batteries.   (Duracell International Inc.)

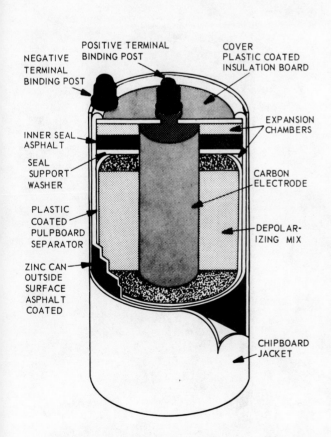

Fig. 5-6. A cutaway view of a No. 6 Dry Cell.

Cells used in flashlights are called PRIMARY CELLS because they generally cannot be recharged. The chemical action is not reversible. However, a cell charger now on the market reportedly can revitalize a cell and increase its useful life.

MERCURY CELLS

Mercury cells are designed to provide high energy and stable voltage. They are used in electronic products ranging from watches to smoke alarms to cardiac pacemakers. Their high energy output permits extensive miniaturization of products that utilize these cells.

Mercury cells are available in sizes ranging from 0.005 to 3.0 cubic inches. The basic chemicals used in their production are a mercury oxide cathode and a zinc anode (compacted powders) and an alkaline electrolyte (liquid).

These cells are made in two general types:

1. Pure mercuric oxide.
2. Mercuric oxide with a small fraction of manganese dioxide.

The first type, at 1.35 volts, maintains extremely stable voltage over a considerable period of time. It is used extensively as a voltage reference and in high reliability applications. The second type of mercury cell, at 1.40 volts, is used in general applications where the need for stable voltage is less critical.

Mercury cells, Fig. 5-7, come in three basic constructions:

1. Flat cells used in hearing aids and watches.
2. Cylindrical cells found in general use.
3. Wound anode cells that feature improved low temperature performance.

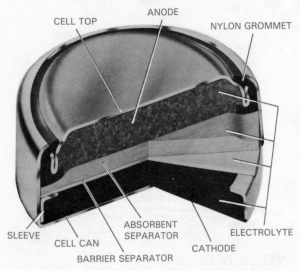

Fig. 5-7. Cutaway drawing of a mercury cell. (Duracell International Inc.)

ALKALINE CELLS

Alkaline manganese cells are used to power a wide range of products where the voltage stability and high energy of the mercury system are not of prime importance. Alkaline cells are used in radios, recorders, cameras and calculators, both in original equipment and in the replacement market.

The basic components of the alkaline system are a manganese dioxide cathode, a zinc anode and an alkaline electrolyte. As with mercury cells, the cathode and anode are compacted powders, while the electrolyte is a liquid. These components are sealed in a steel jacket with a plastic top. End caps and outer steel jackets

complete the assembly. See Fig. 5-8.

Alkaline manganese cells are available in cylindrical and button shapes. The cylindrical cells are in common use in many consumer devices. The flat cells are used in products that require a compact power source.

ACIDS EAT AND BURN

Many batteries, like the automotive battery, will use an acid electrolyte. Drops of this acid can burn your hands and eat holes in your clothing. You must use great care when working around them. Acid can be neutralized with bicarbonate of soda.

THE AUTOMOTIVE BATTERY

Examples of cells which can be recharged many times are the lead-acid cells used in the automotive storage battery. These cells are SECONDARY CELLS. A cutaway battery is il-

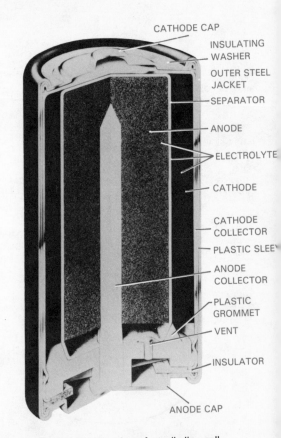

Fig. 5-8. Cutaway drawing of an alkaline cell. (Duracell International Inc.)

Fig. 5-9. A cutaway view of a typical automotive battery.
(ESB Brands, Inc.)

lustrated in Fig. 5-9. Here you can see the electrodes or plates and the construction of the battery. The negative plates are made of spongy lead. The positive plates are lead peroxide. The electrolyte is sulfuric acid and water.

During discharge, the acid goes into the plates, chemically changing them to lead sulfate. The electrolyte becomes nearer to pure water.

A setup for recharging a battery is shown in Fig. 5-10. An external power source is connected so a current is forced through the battery in the opposite direction. This reverses the chemical action. The plates return to their original state. The acid leaves the plates and makes the electrolyte richer in acid.

A convenient way to check the "state of charge"

of the battery is with a HYDROMETER. In a hydrometer, a calibrated floating glass tube sinks less in a liquid heavy with acid than in light liquid. The next time you visit a service station or your school automotive shop, ask the attendant or teacher to show you a HYDROMETER. A similar hydrometer is also used to check the amount of antifreeze in the radiator of your car.

A RECHARGEABLE DRY CELL

In the last few years, many self-powered electric appliances have appeared on the market. These contain their own cells, which can be recharged by plugging into the home power source. These products include flashlights, electric razors, electric toothbrushes, electric carving knives, electric drills and many others.

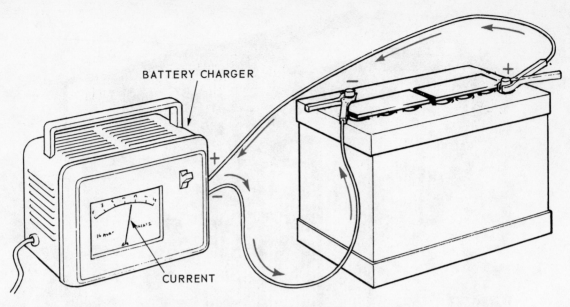

Fig. 5-10. A battery charger connected to battery to show
reverse CURRENT which, in turn, reverses chemical action.

The rechargeable dry cells usually are nickel-cadmium cells with sintered powdered nickel plates formed on a nickel wire screen. The positive plate is treated with nickel salt solutions. The negative plate uses cadmium salts. The electrolyte is potassium hydroxide. An explanation of the chemical action is beyond the scope of this text.

USING CELLS IN SERIES

In most applications, a single cell will not produce high enough voltage to do the required job. Secure a flashlight cell and measure its voltage. It will be about 1.5 volts. When a higher voltage is needed, CELLS are connected in SERIES. Two or more cells connected together make a BATTERY.

The series method of connecting cells is shown in Fig. 5-11. The schematic is also shown so that you will recognize this easy way to show a circuit. Note that now the total output voltage is equal to the voltage of one cell (1.5 volts)

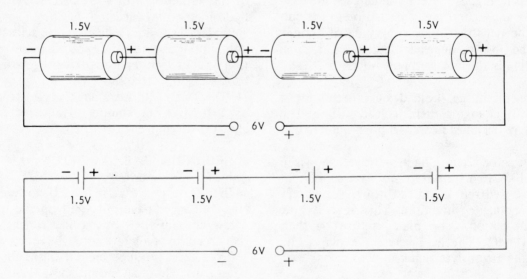

Fig. 5-11. Cells connected in SERIES, NEGATIVE to POSITIVE. Total voltage is 6 V.

times the number of cells (4) or 1.5V × 4 = 6 volts. To connect cells in SERIES, the negative of one cell is connected to the positive of the next, and so on. You might say the cells are connected in line or end to end. A common example of this type of battery is in a 3 cell flashlight. See Fig. 5-12.

Frequently, cells are connected to produce a current over a long period of time, when an increase in voltage is not needed. This is the PARALLEL or SIDE BY SIDE connection. In Fig. 5-13, the four 1.5 volt cells are connected in this manner. The total voltage remains only 1.5 volts. The capacity to supply current has increased about four times. Each cell contributes its part to the total current.

When cells are connected in series, output voltage increases. When in parallel, output voltage is the same as the voltage of one cell.

INTERESTING EXPERIENCES WITH CELLS

Secure four flashlight cells from your school's supply. Measure the voltage of each cell.

1. Connect two cells in series. Measure the voltage.
2. Connect three cells in series. Measure the voltage.
3. Connect four cells in series. Measure the voltage.
4. Connect two cells in parallel. Measure the voltage.
5. Connect three cells in parallel. Measure the voltage.
6. Connect four cells in parallel. Measure the voltage.

Look over an automotive storage battery. If it is the 12-volt type, it will have six cells of 2 volts each in series.

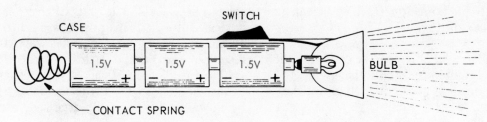

Fig. 5-12. A flashlight using three D cells in SERIES. Total voltage is about 4.5V.

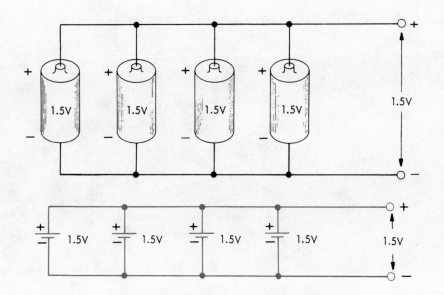

Fig. 5-13. Cells connected in PARALLEL. Terminal voltage remains the same as the voltage of one cell.

ELECTRICAL ENERGY FROM LIGHT ENERGY

One of the more exciting experiences you will have in electronics is the conversion of light to electricity. A device used for this purpose is called a SOLAR CELL. When light strikes the junction between certain semiconductor

In larger scale applications, solar cells are used in various solar energy projects. The Department of Energy is exploring various approaches to developing alternative energy resources for American agriculture. In Nebraska, for example, solar cells power an experimental irrigation project. See Fig. 5-14. The cells turn the sun's radiant energy directly into electricity.

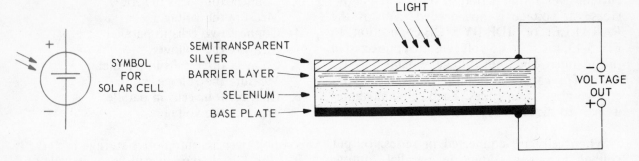

Fig. 5-14. The construction of a photovoltaic cell or SOLAR CELL.

materials, a voltage is produced. A sketch illustrating this phenomenon (fact of scientific interest) appears in Fig. 5-14.

A cell similar to this is used in camera light meters. The brighter the light: the higher the voltage developed. A meter is used to indicate the brightness.

This irrigation project uses 120,000 individual cells to produce 25 kilowatts of electrical power at peak sunlight. This power is used to drive a 10 horsepower pump, that takes water from the reservoir and pumps it through the irrigation system installed in 80 acres of corn and soybeans. See Fig. 5-15. Projects such as this are being watched for possible large scale adoptions.

Fig. 5-15. Solar cells used to produce electrical power.
(U.S. Department of Energy)

BUILDING A SIMPLE LIGHT METER

To build a simple light meter, secure a B2M Solar Cell and connect it to your voltmeter. Follow the diagram, Fig. 5-16. Then, direct a flashlight on the solar cell and observe the movement of the meter.

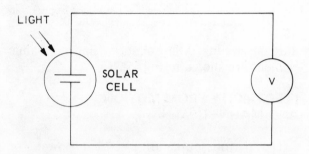

Fig. 5-16. A simple circuit to measure light intensity.

An improved light meter, Fig. 5-17, uses a battery and a transistor to increase its sensitivity to light change. You will study transistors later in this text.

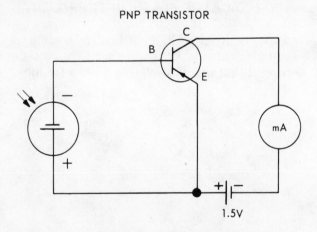

Fig. 5-17. An improved light meter using a TRANSISTOR to increase its sensitivity.

PHOTOCONDUCTIVE CELLS

Another component that takes advantage of light intensity is a cell which CHANGES ITS RESISTANCE when exposed to light. A sketch of such a device and its symbol are shown in Fig. 5-18. A photoconductive material (cadmium sulfide) is deposited on an insulating disc with a conducting layer deposited over the cadmium sulfide. When light shines on this

material, it will increase its conductivity or decrease its resistance. Note that this cell controls the current in a circuit and can be used directly to operate meters and relays.

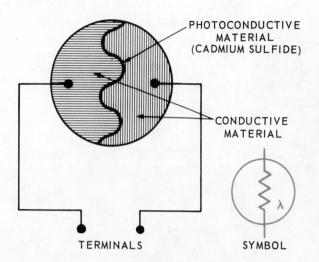

Fig. 5-18. The construction and electronic symbol for a Cadmium-Sulfide photoconductive cell.

CHANGING RESISTANCE BY LIGHT

In the circuit in Fig. 5-19, the photoconductive cell is connected in series with the current meter and a power source. As the intensity of the light increases, the current also increases. Such devices are widely used in industry for counting and sorting, fire and burglar alarms, automatic door openers, safety warning systems and for reading information into computers. Some interesting circuits will be described in later Units.

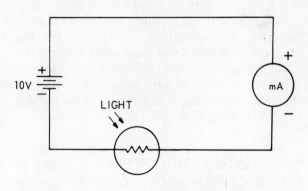

Fig. 5-19. A light sensitive circuit using the photoconductive cell. The brighter the light, the higher the current.

ELECTRICITY FROM HEAT

When two dissimilar metals are joined together and heated, a slight potential difference (voltage) is produced. A device that makes use of this effect is known as a THERMOCOUPLE. To demonstrate this means of producing voltage, perform the experiment illustrated in Fig. 5-20. Note that the dissimilar wires used in this experiment are iron and copper.

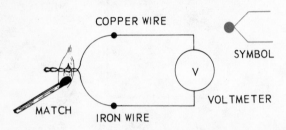

Fig. 5-20. A voltage is produced by heating the THERMOCOUPLE.

The commercial type of thermocouple may use wires of copper-constantan, iron-constantan or platinum-platinum rhodium. Constantan is an alloy of about 40 percent nickel and 60 percent copper. The thermocouple is used industrially for temperature control and monitoring (observing and checking). If a higher output voltage is required, several thermocouples may be connected together to form a THERMOPILE.

ELECTRICITY FROM MECHANICAL PRESSURE

A very interesting phenomenon occurs when you bend or apply pressure to crystals such as quartz, tourmaline and certain ceramic materials. Actually, the distortion causes the electrons to move out of their usual orbits and a small voltage is produced between the sides or surfaces of the crystal. When the force is removed, the electrons return to their normal positions. See Fig. 5-21.

A laboratory apparatus for demonstrating this PIEZOELECTRIC EFFECT is shown in Fig. 5-22. Applications that utilize this unusual effect are crystal microphones, crystal cartridges for record players and many other in-

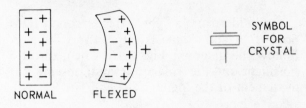

Fig. 5-21. When a crystal is bent or flexed, the electrons are distorted out of their normal positions.

dustrial circuits. Microphones using a ceramic element are shown in Fig. 5-23.

ELECTRICITY FROM MOTION AND MAGNETISM

The major source of electrical power for home and industry is found in the reaction between a moving conductor and a magnetic field. The GENERATOR, for example, makes use of this reaction. In Unit 8, we will study this most important principle of producing electrical energy.

LED INDICATOR FLASHLIGHT PROJECT

Have you ever had difficulty finding a flashlight in the dark? If so, you will appreciate a project that will allow you to build a flashing,

Fig. 5-22. A laboratory demonstrator for piezoelectricity. When the crystal is flexed by moving the lever, the generated electricity will light the lamp. (Central Scientific Co.)

Fig. 5-23. Typical ceramic type microphones used with citizens band, amateur, fixed station and mobile radio transmitters. (Shure Bros., Inc.)

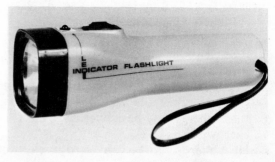

Fig. 5-24. Light Emitting Diode (LED) Indicator Flashlight.

light emitting diode (LED) circuit into a flashlight. See Fig. 5-24. The two cells (3 volts) in the flashlight power the LED circuit, which causes very little power drain.

See the LED identification in Fig. 5-25. Refer to Fig. 5-26 for the simple schematic used for this circuit. Construct the circuit. Then, mount it anywhere in the flashlight case, but watch out for "shorts" as you install the circuit.

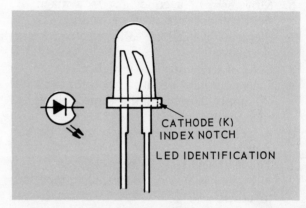

CATHODE (K) INDEX NOTCH

LED IDENTIFICATION

Fig. 5-25. LED identification. Note location of notch.

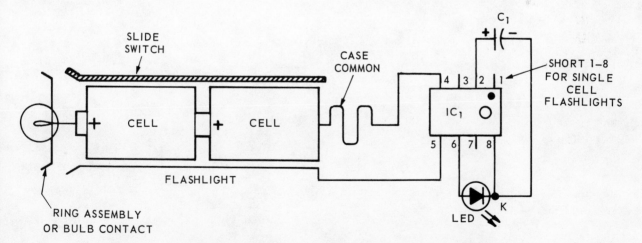

PARTS LIST FOR LED INDICATOR FLASHLIGHT

C_1 — electrolytic capacitor, 300 μF @ 3 WVdc

IC_1 — integrated circuit, National Semiconductor LM 3909

LED — 1 1/2V light emitting diode, NSL 5027 (National Semiconductor or equivalent)

Flashlight — 2 cell unit with sufficient room to install circuit

Misc. — wire, solder, IC socket

Fig. 5-26. Schematic and parts list for LED Indicator Flashlight.

FORWARD STEPS IN UNDERSTANDING ELECTRICITY-ELECTRONICS

1. Electricity is produced by converting a form of energy to electrical energy.
2. Chemical energy may be converted to electrical energy.
3. Light energy may be converted to electrical energy.
4. Heat energy may be converted to electrical energy.
5. Mechanical energy may be converted to electrical energy.
6. Movement of a conductor in a magnetic field will produce electrical energy.
7. Elements in a cell are called electrodes and electrolyte.
8. Cells in series produce a higher voltage.
9. Cells in parallel produce a greater capacity to supply current of the same voltage as each individual cell.
10. A primary cell generally cannot be recharged.
11. The lead-acid cell can be recharged.
12. A battery is two or more cells connected together.
13. Light energy will change the resistance of certain kinds of material.
14. Electrical energy may be produced from heat energy by using a thermocouple.
15. Piezoelectricity is produced by pressure or by flexing a crystal.

TEST YOUR KNOWLEDGE - UNIT 5

1. Six volts are required to operate a small lamp. You have several flashlight cells on hand. Draw the circuit and connect the cells to produce the required brightness.
2. Connect two D cells in series to the 6 volt lamp. Does the lamp glow at full brightness? Explain.
3. Using the ohmmeter, what is the resistance of the 6 volt lamp?
4. Did the lamp glow when you connected the ohmmeter to it? Why?
5. Design a circuit which would electronically tell you if it was day or night. Could you use the same circuit to tell if it is a sunny or cloudy day?
6. Design a circuit using a photoconductive cell by which you can tell if it is night or day.

Unit 6

CONTROLLING ELECTRICITY

The principles of electricity and electronics are based on the theory that electrons can be controlled.

Major concepts in this Unit are:

1. What a conductor's size and type have to do with its ability to pass electrons.
2. What an electrical circuit is.
3. How a series circuit works.
4. What a parallel circuit is.
5. When a circuit is considered to be "open" or "shorted."
6. How fuses and circuit breakers operate.
7. What the basic types of switches are.

CONDUCTORS AND CURRENT

When we have a potential difference or voltage, we have a source of electrical energy. Through experiences in earlier Units, you learned that it is necessary to connect this voltage to some device or load to produce work or some desirable effect. The connections are made by wires or conductors. A CONDUCTOR is a low resistance path or road for electron current to flow.

It might be well at this time to understand exactly how current flows in a wire. It does NOT run through wire like water runs through a pipe.

Fig. 6-1 is an enlarged view of a piece of copper wire. Only a few atoms of copper are shown in the drawing. In a given wire, there are billions of such atoms. As a negative potential is applied to one end of the wire, the force or energy is transferred from electron to electron to the positive end of the wire or source. At this positive end, electrons are attracted off the wire.

These are facts you should remember:

1. At no time does the conductor store electrons. It gives up one electron at the positive end for each electron forced on at the negative end.
2. The ability of a conductor to conduct depends on the number of free electrons available for conduction.
3. The actual drift of electrons through the wire is relatively slow.
4. The transfer of electrical energy approaches the SPEED OF LIGHT, which has been measured as 186,000 miles per second or 300,000,000 meters per second.

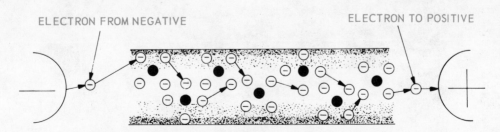

ELECTRON FROM NEGATIVE   ELECTRON TO POSITIVE

Fig. 6-1. The DRIFT of free electrons through a conductor is a current.

In the physics laboratory, the transfer of energy may be demonstrated by a line of balls in contact with each other. Refer to Fig. 6-2. When the first ball rolls down and hits the next ball, the energy is transferred through the balls and the last or eighth ball is forced away from the string.

TYPES AND SIZES OF CONDUCTORS

Hundreds of types and sizes of wire and cables are required to meet the needs of the electronics industry. You have already learned that the size of wire required is determined by the amount of current it must conduct. A small wire has more resistance than a large wire.

A study of the table in Fig. 6-3 will show that

SERIES OF BALLS IN CONTACT WITH EACH OTHER

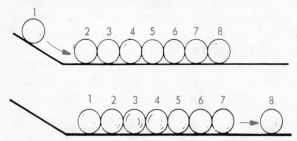

Fig. 6-2. This experiment demonstrates that a force applied to one end of a string of balls is transferred through the balls to the opposite end.

wire sizes are identified by gage number. The larger the number, the smaller the wire. Also note the resistance per 1000 ft. of each size. Fig. 6-4 shows some common types of wire you will use in this course.

ELECTRIC CIRCUITS

Electric and electronic circuits may be divided into FOUR parts. Each part is identified by its use. An awareness of these parts of the total system will help you understand how they work and how to service the equipment.

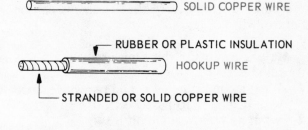

Fig. 6-4. Three common types of conductors.

GAGE NO.	DIAM. MILS.	OHMS PER 1,000 FT. OF COPPER WIRE AT 25°C.	GAGE NO.	DIAM. MILS.	OHMS PER 1,000 FT. OF COPPER WIRE AT 25°C.
1	289.3	0.1264	21	28.46	13.05
2	257.6	0.1593	22	25.35	16.46
3	229.4	0.2009	23	25.57	20.76
4	204.3	0.2533	24	20.10	26.17
5	181.9	0.3195	25	17.90	33.00
6	162.0	0.4028	26	15.94	41.62
7	144.3	0.5080	27	14.20	52.48
8	128.5	0.6405	28	12.64	66.17
9	114.4	0.8077	29	11.26	83.44
10	101.9	1.018	30	10.03	105.2
11	90.74	1.284	31	8.928	132.7
12	80.81	1.619	32	7.950	167.3
13	71.96	2.042	33	7.080	211.0
14	64.08	2.575	34	6.305	266.0
15	57.07	3.247	35	5.615	335.0
16	50.82	4.094	36	5.000	423.0
17	45.26	5.163	37	4.453	533.4
18	40.30	6.510	38	3.965	672.6
19	35.89	8.210	39	3.531	848.1
20	31.96	10.35	40	3.145	1069.

Fig. 6-3. Wire Gage and Resistance chart.

1. The POWER SOURCE.
2. The CONDUCTORS or CURRENT PATHS.
3. The CONTROL.
4. The LOAD or WORK to be done.

In Fig. 6-5, each part is identified in a simple circuit. Fig. 6-6 shows the schematic diagram of the same circuit.

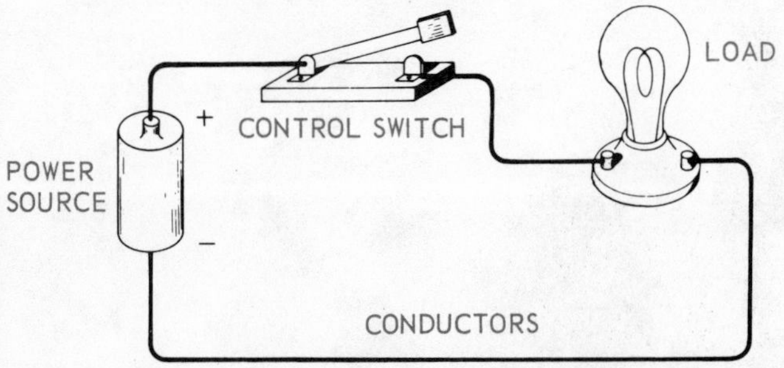

Fig. 6-5. The basic electrical circuit with power source, conductors, control and load.

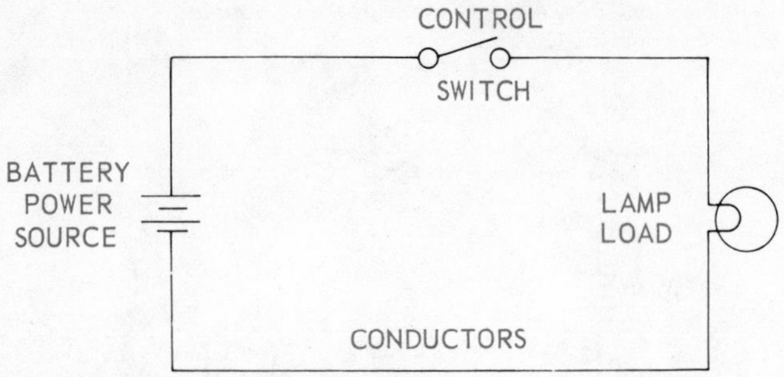

Fig. 6-6. The schematic diagram of the basic electrical circuit.

The parts of the basic circuit may take many forms. The SOURCE OF ENERGY may be any one of the five sources of energy. The LOAD may be a lamp, a motor, a heater, a radio, a TV or any one of hundreds of other devices. The CONTROL may be a simple "on and off" switch or any number of devices which would limit the current or the voltage of the circuit.

SERIES CIRCUIT

In a SERIES circuit, there is only one path for the current to flow, and the same current flows through each and every component in the circuit.

In Fig. 6-7, a 6 volt battery is connected to:

A. Two lamps in series.
B. Three lamps in series.
C. Four lamps in series.

The path of current flow is marked by arrows in each circuit.

EXPERIENCES WITH SERIES CIRCUITS

1. Connect a single lamp in the circuit, as in Fig. 6-8. Observe how brightly the lamp glows.
2. Connect the milliammeter in series with the circuit and measure the current required to light the lamp. The circuit in Fig. 6-8 includes a voltmeter, milliammeter and a single-pole switch. A student laboratory set-up for this experience is shown in Fig. 6-9.

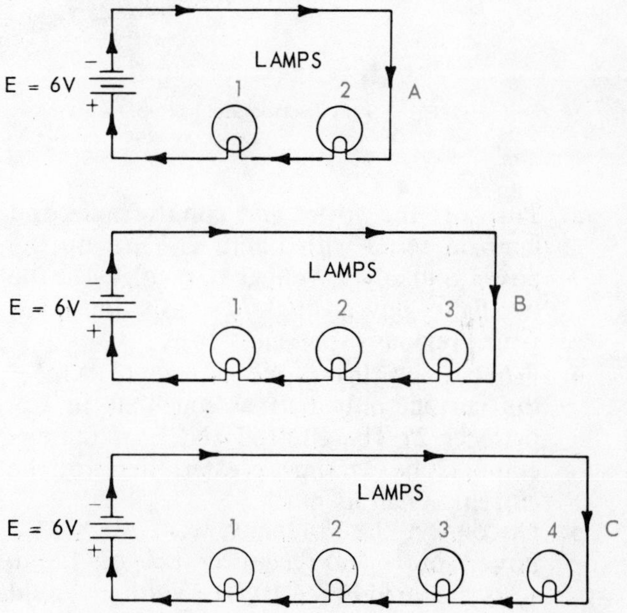

Fig. 6-7. Two, three and four lamps connected in SERIES.

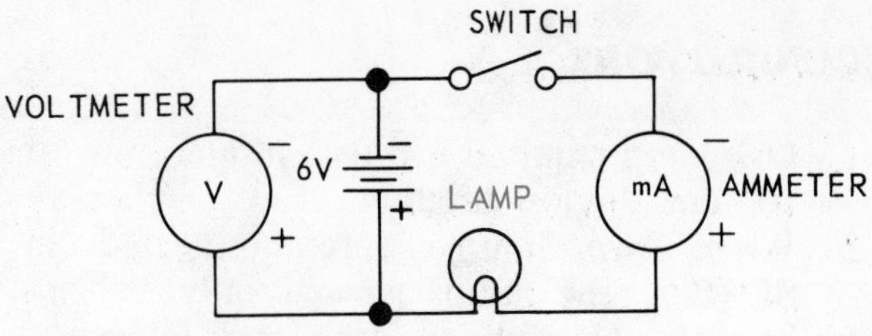

Fig. 6-8. The circuit used in the text to measure current and voltage in a simple one lamp circuit.

Fig. 6-9. In this laboratory setup, two lamps are connected in series. A meter measures the source voltage.   (Lab—Volt, Buck Engineering Co.)

3. Turn off the power and connect a second lamp in series with Lamp 1. Turn on the power and adjust voltage to 6 volts. Do the two lights glow as brightly as only one did in your previous experience? Why?

4. What is the current measurement? Why is the current only half as much as in Experience 2? The RESISTANCE of the circuit must be two times greater, therefore the current is half as much.

5. Disconnect the voltmeter from across the power source and reconnect it across Lamp 1, as shown in Fig. 6-10. The voltage should read about 3 volts. Connect the voltmeter across Lamp 2. It also should read about 3 volts.

CONCLUSIONS

1. One lamp required 6 volts and a current of 100 mA to glow brightly.

2. When two lamps were connected in SERIES, the lamps glowed only half as brightly. The voltage across each lamp was only 3 volts and the circuit current was half as much.

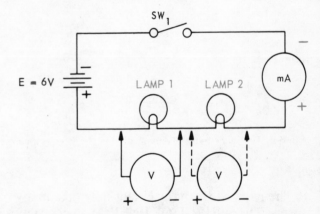

Fig. 6-10. This circuit illustrates method of measuring voltage drops across the lamps.

3. By adding a second lamp in SERIES, the circuit resistance must have doubled. The total resistance of the circuit must be:

R Lamp 1 + R Lamp 2 = TOTAL R

4. Voltage across Lamp 1 plus voltage across Lamp 2 is equal to source voltage or:

3V + 3V = 6V (source voltage)

5. The voltage loss across either lamp is called the "voltage drop," and the sum of the voltage drops equals the source voltage.
6. The voltage drop across a component can always be found by the formula of current times resistance:

I Circuit × R Component = Voltage Drop

$$I \times R = E$$

The voltage drop is frequently called the IR drop.

Repeat the experiences with SERIES CIRCUITS, using 3 lamps and 4 lamps until you are satisfied that you completely understand these conclusions. These experiences are among the most important lessons in your study of electricity.

PARALLEL CIRCUITS

When components are connected, side by side, so that there is more than one path for current flow, the components are connected in PARALLEL. In Fig. 6-11, a six volt power source is connected to:
A. Lamp 1 only.
B. Lamp 1 in parallel with Lamp 2.
C. Lamp 1, Lamp 2, Lamp 3 in parallel.
D. Lamp 1, Lamp 2, Lamp 3 and Lamp 4 in parallel.
The paths for the current flow for the lamps are indicated by arrows.

EXPERIENCES WITH PARALLEL CIRCUITS

1. Connect the circuit with power source, milliammeter and voltmeter and one lamp as displayed in schematic drawing Fig. 6-12. Does the lamp glow brightly? What is the voltage across the lamp? How much current does this one lamp use?
2. In Fig. 6-13, Lamp 2 is connected in parallel with Lamp 1. Does Lamp 2 glow as brightly as Lamp 1? What is the voltage across L_1? What is the voltage across L_2? Are these voltages the same? But look at the MILLIAMMETER! The current reads twice as much as the single lamp circuit.
    Total current drawn from the source must equal current of L_1 plus current of L_2.

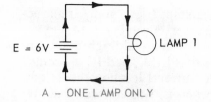
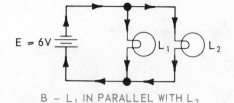
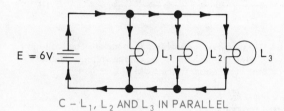
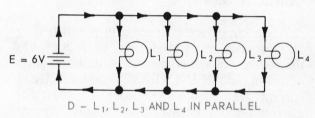
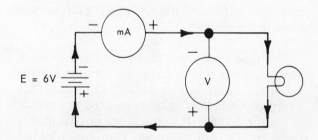

Fig. 6-11. One, two, three and four lamps in PARALLEL.

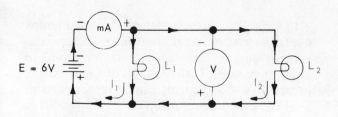

Fig. 6-12. A single lamp connected to voltage source with voltmeter measuring voltage across lamp and current meter to measure TOTAL CURRENT drawn from source.

Fig. 6-13. Two lamps connected in parallel. The voltage across either lamp is the same as the source. The total current is equal to the SUM of current I_1 and I_2.

3. Add the third Lamp, L₃, to your circuit according to the schematic in Fig. 6-14. Do L₁, L₂ and L₃ burn with equal brightness? Are the voltages across L₁, L₂ and L₃ the same? But look at the current! It is now three times greater than when L₁ was connected alone.

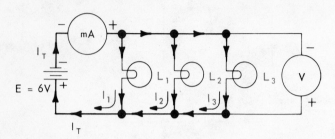

Fig. 6-14. Three lamps connected in parallel. The voltage across each lamp is the same as the source voltage. The total current, I_T, is equal to the SUM of $I_1 + I_2 + I_3$.

CONCLUSIONS

When lamps are connected in parallel, they all glow with equal brightness. Each lamp has the source voltage connected to it. Each lamp added in parallel INCREASES the total circuit current drawn from the source. In order for current to increase, the total circuit RESISTANCE MUST DECREASE. Each lamp added to the circuit in parallel will decrease the total resistance, and the circuit will draw more current.

Continue your experience using four lamps in parallel. Your power company supplies a fixed voltage to your home, usually 117 volts. Are your lights and appliances connected in series or in parallel? Why? If you remove one lamp in a series circuit, will the others continue to glow? Why? If you remove one lamp from a parallel circuit, will the others continue to glow? Why?

WHAT IS A LOAD?

In your previous experience, you used lamps to place a load on the power source. Other components and devices can also be connected in series or parallel. If they use POWER from the source these components can be represented as a resistive load (resistance).

Resistance is the only property of a circuit which uses POWER. The circuits previously

used for series and parallel experiences are redrawn in Figs. 6-15 and 6-16. The lamps are replaced with resistors, each with a value equal to the resistance of the lamp it represents.

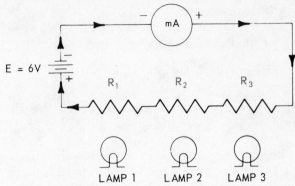

Fig. 6-15. The series circuit where L₁, L₂ and L₃ have been replaced with their equal resistances R₁, R₂ and R₃.

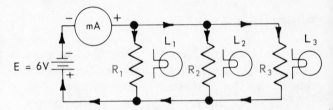

Fig. 6-16. The parallel circuit where L₁, L₂ and L₃ have been replaced with their equal resistances R₁, R₂ and R₃.

WHAT IS A SHORT CIRCUIT?

Water flowing down a mountain will take the easiest path. Electricity also follows this law of nature. It will always flow in the PATH OF LEAST RESISTANCE. Looking at Fig. 6-17, a power source is connected to a resistive load. The load limits the current. Then, the uninsulated connecting wires are accidentally brought in contact with each other. ZOWIE! Sparks fly and the wires get red hot. This is a SHORT CIRCUIT, a shorted path for the current to get around the load. This dangerous condition can cause severe burns and may set a house on fire! How many lamps and cords like the one in Fig. 6-18 do you have around your house?

Conducting wires are covered with insulating materials such as rubber and plastic. The insulation prevents unwanted and dangerous SHORT circuits between wires and components. A

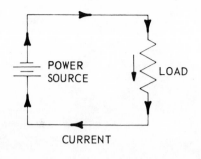

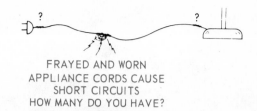

FRAYED AND WORN
APPLIANCE CORDS CAUSE
SHORT CIRCUITS
HOW MANY DO YOU HAVE?

Fig. 6-18. Frayed and worn appliance cords are a serious source of danger and fire!

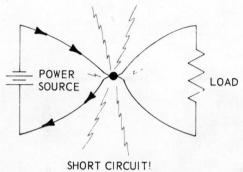

Fig. 6-17. When uninsulated conducting wires touch, the resistance is ZERO. High currents - sparks - and heat result. This is a DANGEROUS CONDITION.

"short" may be tested by using an ohmmeter, which will read zero ohms.

PROTECT YOURSELF AND YOUR PROPERTY

A FUSE is a protective device in an electrical circuit. You should be aware of WHY the fuse is placed in the circuit. As you have previously studied, a SHORT CIRCUIT will cause very high currents to flow from the power source. These currents produce HEAT! The wires can

get RED HOT and even MELT. Any material such as wood or paper nearby can be set on fire! At this very moment, a home belonging to a family like yours is burning to the ground because of improper or insufficient circuit protection. This can be prevented by proper fusing.

Fig. 6-19 shows a variety of fuses used in electrical and electronic circuits.

A fuse is an automatic circuit protector. Common types are made of a short length of metal or wire which quickly melts and "opens" the circuit if more than a certain amount of current flows. See Fig. 6-20, and the symbol for a fuse. Note that the fuse is connected in SERIES with the load. If the load should draw too much current, the fuse element quickly melts and the circuit no longer works. The circuit will not work again until the fuse has been replaced with a new one. BEFORE A FUSE IS REPLACED, THE REASON THAT MADE THE FUSE BURN OUT MUST BE DISCOVERED AND CORRECTED.

When the power company brings electricity to your home, it is connected to an entrance

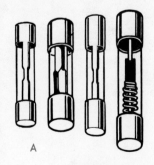

Fig. 6-19. Various types and sizes of FUSES. A—Glass tube fuses used in electronic equipment and automotive electrical circuits. B—Cartridge fuses found in home and industrial electrical circuits. C—Typical plug fuse found in many home electrical circuits. D—The FUSETRON, which has a delay time before blowing. This allows high currents required to start a motor to flow momentarily. (Bussmann Mfg. Div., McGraw Edison Co.)

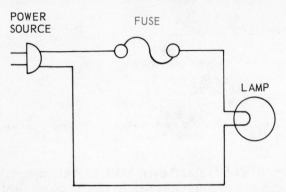

Fig. 6-20. A simple circuit of power source, lamp and protective fuse. In case of a short circuit, the fuse would "blow out."

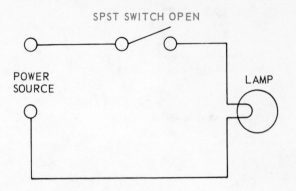

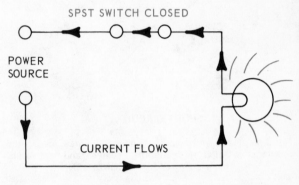

Fig. 6-22. A single-pole switch is used to control the current to the lamp. A SPST switch.

switch and fuse box. Then, the electricity is divided into several circuits to supply the lamps, convenience outlets, heaters, ranges and other devices. Each circuit will have its own individual protection. Some circuits used for lamps only will be protected with 15 ampere fuses. A circuit for kitchen appliances may have a 20 ampere fuse. Other circuits for ranges and heaters will have larger fuses.

The typical ENTRANCE SWITCH and DISTRIBUTION BOX pictured in Fig. 6-21 uses circuit breakers instead of fuses. A CIRCUIT BREAKER is either a thermal or magnetic switch that opens the circuit when too much current flows.

Fig 6-21. A 100 ampere, 120/240 volt main circuit breaker load center. The indicator light shows which circuit breaker is open. (Square D Co.)

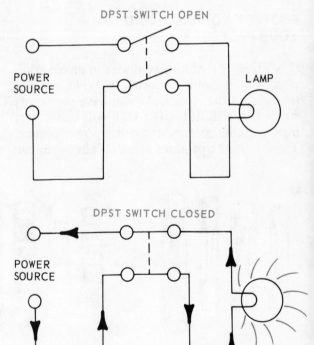

Fig. 6-23. A double-pole switch is used to control the current in both wires leading to the lamp. A DPST switch.

SWITCHES

A SWITCH is a simple component which will OPEN or CLOSE a circuit. A switch used with a lamp will turn the lamp ON or OFF. In the open position, there is no electrical connection between its terminals. In this case, the switch would represent infinity (unlimited) resistance. In its closed position, the contacts are touching each other and there is a path of zero resistance through the switch.

Fig. 6-22 shows a simple SINGLE-POLE switch used to turn the lamp ON or OFF. In Fig. 6-23, a DOUBLE-POLE switch performs exactly the same job. However, the double-pole switch OPENS and CLOSES both wires leading to the lamp. This type of hookup is required under certain circumstances.

In Fig. 6-24, a SINGLE-POLE DOUBLE-THROW (SPDT) switch is used. In position 1, Lamp 1 will glow. In position 2, Lamp 2 will glow. It is an EITHER - OR situation. Both lights cannot glow at the same time. Observe the current flow indicated by arrows.

Fig. 6-25 illustrates a DOUBLE-POLE DOUBLE-THROW (DPDT) switch. The

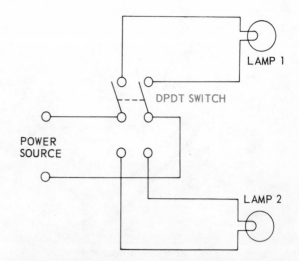

Fig. 6-25. A Double-Pole Double-Throw (DPDT) switch is used to switch both sides of the power line to either Lamp 1 or Lamp 2.

DPDT switch could also be used if both sides of the line required switching.

ROTARY SWITCHES

Rotary switches, Fig. 6-26, are used in many circuits. A rotary switch is a multiple contact

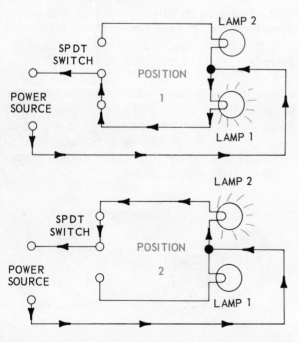

Fig. 6-24. A Single-Pole Double-Throw (SPDT) switch is used to turn either Lamp 1 or Lamp 2 on. Trace the current flow.

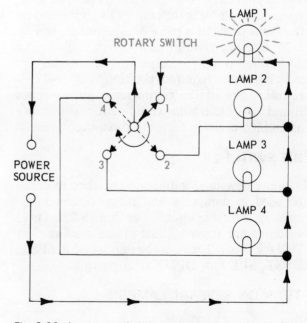

Fig. 6-26. A rotary switch is used to turn on any desired lamp. The current for position 1 to cause Lamp 1 to glow is indicated by arrows. Other positions of switch are shown by dashed lines.

Fig. 6-27. A rotary switch with two wafers of contacts controlled by a single rotary shaft. Many of these types of switches are used in electronic equipment. (Centralab)

switch that may be turned to any desired position. An example is the control knob located on the front panel of a piece of equipment. In Fig. 6-26, a four position rotary switch is used. The desired lamp may be turned on by rotating the switch to the required position. Fig. 6-27 illustrates one of the commercial type switches found in electronic equipment. These are assembled in many forms to meet circuit needs.

DIP SWITCHES

Some new microminiature switches that can be used in computer and microcomputer circuits is the "dipswitch." See Fig. 6-28. These switches are termed "dual-in-line package" or DIP for short. They may be purchased in SPST, DPST, SPDT or DPDT arrangements.

TURN ON SEVERAL LAMPS

A load can consist of one or more lamps in either series or parallel. Since we wish all lamps to glow with full brightness, it will be necessary to connect extra lamps in parallel with the first lamp. This may be observed in Fig. 6-29 where three lamps are turned ON or OFF at the same time by a single switch. What would result if the three lamps were connected in SERIES?

Results: Each lamp would burn at only one third of its required brightness. Also, if one lamp should burn out, all the lights would go out. You may draw your own conclusions as to the method of connecting lamps and appliances in your home wiring circuits.

Fig. 6-28. Dual-in-line package (DIP) switch. (AMP Inc.)

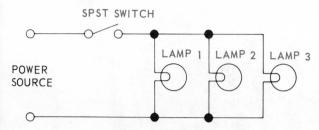

Fig. 6-29. Three lamps in parallel are controlled by one SPST switch. All lamps burn at full brightness.

THREE-WAY SWITCHING

Three-way switching is studied because of its popularity in house wiring circuits. Many times you may wish to turn a light ON as you enter a room and turn it OFF as you leave the room from another door. You may wish to turn lights ON or OFF at either the top or bottom of a staircase. It certainly saves a lot of steps and "wear and tear" on the carpet, as well as providing personal safety by wiring light circuits so that they may be switched from more than one location.

Fig. 6-30 shows the connections for the three-way circuit. Two switches are required. The light can be turned either ON or OFF by either switch. A sketch of a typical three-way switch

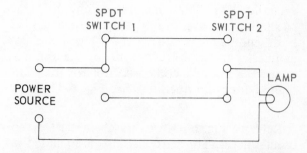

Fig. 6-30. Three-way switch connections. The lamp can be turned "on or off" by either switch 1 or 2.

connection used in house wiring appears in Fig. 6-31. Note the three conductor cable is used from switches to junction box. Power is brought to the junction box by a two wire cable. The lamp is shown in the ON position and the complete current path is marked by colored arrows.

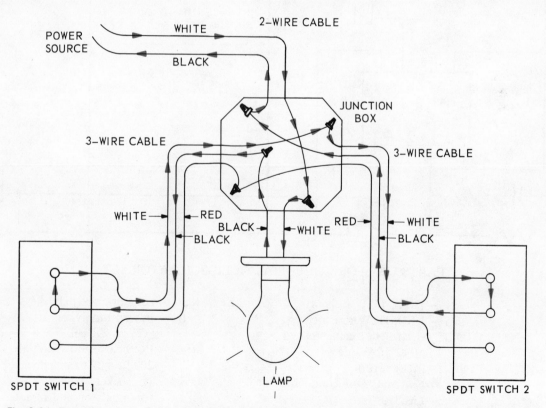

Fig. 6-31. Typical connections in home wiring for three-way switches to operate one overhead lamp. Follow current path marked with arrows. Lamp is ON. Note wire color code and how wires are connected. Compare with Fig. 6-30.

CODE PRACTICE OSCILLATOR PROJECT

A simple, yet effective, code practice oscillator can be built as a project, using the versatile LM 3909 integrated circuit. One oscillator simultaneously drives speakers at both sending and receiving ends. These two sets can be separated up to 200 feet or the unit can be used as a simple code oscillator for the practicing ham radio operator. See Fig. 6-32.

Refer to Fig. 6-33 for the schematic for this oscillator set. The parts list also is included in the illustration. Note that a power supply or a 1 1/2V "C" or "D" cell may be used.

The three-way system and parallel telegraph keys (Key$_1$ and Key$_2$) allow beginners to use the set without having to understand the use of a "send-receive" switch. Have fun!

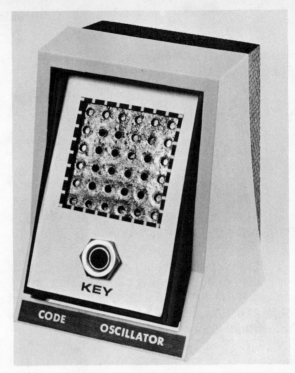

Fig. 6-32. Code Practice Oscillator.

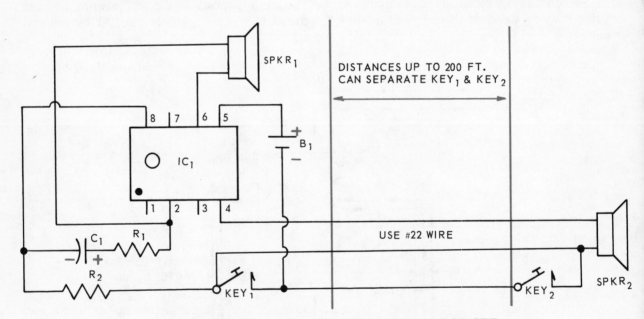

PARTS LIST FOR CODE PRACTICE OSCILLATOR SET

R$_1$ — 330 Ω, 1/2W resistor
R$_2$ — 1 KΩ, 1/2W resistor
C$_1$ — electrolytic capacitor, 1 μF @ 16 WVdc
IC$_1$ — integrated circuit, National Semiconductor LM 3909
SPKR$_1$, SPKR$_2$ — 2 in., 8 Ω speakers

Key$_1$, Key$_2$ — code practice hand keys
B$_1$ — power supply or 1 1/2V "C" or "D" cell
Misc. — PC board and materials, 200 ft. #22 stranded wire, IC socket, solder, cases for oscillator and remote speaker, decals

Fig. 6-33. Schematic and parts list for Code Practice Oscillator Set.

FORWARD STEPS IN UNDERSTANDING ELECTRICITY-ELECTRONICS

1. An electrical circuit consists of a power source, a load, the control and connecting wires.
2. Current flow in a circuit is the transfer of energy between electrons in a conductor.
3. A poor or nonconductor is called an insulator.
4. Wires are sized by Gage Number, the larger the number, the smaller the wire.
5. Electrical energy travels at approximately the speed of light or 186,000 miles per second or 300,000,000 meters per second.
6. In a series circuit, there is only one path of current flow, and the current must be the same at any point in the circuit.
7. Series connections of resistive components, such as lamps, increase the total resistance of the circuit.
8. In a series circuit, the voltage divides according to the resistance of the component and the current. It is called the IR drop.
9. In a parallel circuit, there are multiple paths for current flow. Components are connected side-by-side.
10. In a parallel circuit, the total current is equal to the sum of all the branch currents.
11. Adding more resistive components in parallel increases the total circuit current and, therefore, must also reduce the total resistance.
12. Only resistance as a load uses power from the source.
13. A short circuit is zero resistance, and maximum current flows.
14. A fuse is a circuit protective device.
15. A switch is a control component used to open or close a circuit. Open means to TURN OFF. Close means TURN ON.

RULES OF SAFETY

OCTOPUS EXTENSIONS

Older homes frequently have too few wall receptacles to serve all the lamps and appliances required for modern daily living. The use of multiple extension cords to a single receptacle is dangerous. Wires may get hot and burn! You may get some severe shocks!

FUSES PROTECT

A circuit designed to carry a certain current should have a fuse which will BLOW OUT if the circuit current is exceeded. Always look for the cause that burned out the fuse. Do not replace a fuse until the cause of its BLOW OUT is found. Never replace the blown fuse with a larger fuse or a jumper. You are inviting trouble.

FIRST AID

All well-run shops and laboratories have FIRST AID kits in handy locations. Know where the first aid kit is in your shop. Report any accident, no matter how minor. Your teacher will supply treatment where needed.

SAVE YOUR BACK

Some electronic equipment is heavy, especially the types which use transformers. Always get help to lift it or move it around. Ask your teacher to demonstrate the proper way to lift a heavy object.

HOT SOLDERING IRONS

A hot soldering iron or gun appears innocent. If it is left on the bench or hung in the tool cabinet, some unsuspecting fellow student may put a hand on it. DO NOT let this happen.

LONG CORDS DROP VOLTAGE

The longer the extension cord, the greater the resistance, and the voltage at its end may be somewhat less than required for an appliance or motor. The life of your tools and appliances may be shortened by operating them with reduced voltage.

SAVE YOUR SIGHT

Your eyesight is a valuable gift of life. Protect your eyes when grinding, chipping or polishing. Also see that students nearby are protected.

TEST YOUR KNOWLEDGE - UNIT 6

1. What is the resistance in ohms of 100 feet of No. 20 hookup wire?
2. Measure the resistance of a foot of No. 20 Nichrome wire.
3. What would happen if you used Nichrome wire to connect up a lamp circuit?
4. In a series circuit, why do all lamps go out when one lamp is removed from its socket?
5. Why would the electrician wire your home with several independent circuits, rather than one heavy circuit?
6. All outlet plugs and ceiling lights in one room of a house should be on the same circuit. Yes or No? Why?
7. Which will use more electric power? Four lamps in series or four lamps in parallel? Explain.
8. Draw the wiring diagram for problem 7 and then prove circuit by actual connections.

    EXPERIENCE A. Connect Lamp 1 only to be controlled by single-pole switch.

    EXPERIENCE B. Connect Lamps 1, 2 and 3 to be turned on by single-pole switch. All lamps to burn brightly.

    EXPERIENCE C. Connect Lamp 1 in a three-way switch circuit so that it may be controlled by either SW₁ or SW₂.

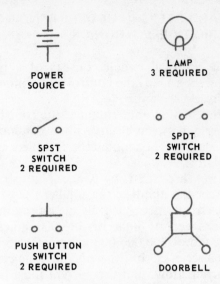

POWER SOURCE

LAMP
3 REQUIRED

SPST SWITCH
2 REQUIRED

SPDT SWITCH
2 REQUIRED

PUSH BUTTON SWITCH
2 REQUIRED

DOORBELL

USE THESE COMPONENTS IN THE FOLLOWING EXPERIENCES.

EXPERIENCE D. Connect circuit with SPDT switch to light Lamp 1 in one position and Lamp 2 in its other position.

9. Draw a wiring diagram of a doorbell circuit for your home. Use push buttons for front and rear doors to ring bell.
10. Draw a circuit using 3 eight-pole rotary switches to operate a lamp or special magnetic lock. The secret combination to activate the lamp or lock will be 3-5-7.

Unit 7

THE MYSTERY OF MAGNETISM

Magnets are very important devices in the field of electricity. In this Unit, you will learn:

1. What magnetic fields are.
2. What the laws of magnetism are.
3. How the Left Hand Rule applies to electromagnetism.
4. When an electromagnet becomes a solenoid.
5. How an electromagnet operates.
6. How a relay works.
7. What a reed relay is.

THE NATURAL MAGNET

Legends handed down for centuries told of a mysterious kind of stone found on the earth's surface in Asia Minor. This stone attracted small pieces of iron, and it was not until the 12th century that English scientists decided this suspicious stone did not contain life. Rather, in some manner, it was related to another mysterious force called "electricity."

In the early 17th century, Sir William Gilbert proposed the almost unbelievable theory that the earth on which we live is an enormous magnet. He found that a small living (magnetic) stone suspended on a string or placed on a small wooden float in a vessel of water would turn in a NORTH-SOUTH direction. Early navigators realized the value of the living stone and used it as a direction finder when sailing their ships in unknown seas.

The natural magnet has been replaced in the electronic age by more dependable and stronger permanent magnets made of steel and alloys. Much has been learned about the magnetic field of force existing in space around the surface of the earth.

VISIBLE MAGNETIC FIELDS

Although you cannot look into space and see a magnetic field, such a field does exist in space all about us. The effect of this field can be seen by performing the learning experience shown in Figs. 7-1 and 7-2. Place a sheet of cardboard over a single magnet. Then, sprinkle fine iron

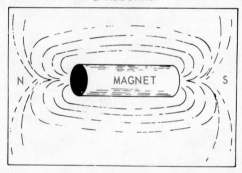

Fig. 7-1. Iron filings arrange themselves to form a pattern of the magnetic field.

Fig. 7-2. The magnetic field can be seen and photographed by using this demonstration device.
(Lab—Volt, Buck Engineering Co.)

filings on the cardboard, much like you would sprinkle salt on a hamburger. Now, gently tap the cardboard and the iron filings will arrange themselves along the "lines of magnetic force" between the north and south poles.

Some observations should be made while studying these lines of force:
1. Each line is an individual line and no two lines cross each other.
2. The lines are dense at each pole and enter the pole at a perpendicular angle.
3. Each line forms a continuous path from north to south and through the magnet to north. There is no movement within this path.

LAWS OF MAGNETISM

Place two small magnets close together. If UNLIKE POLES are next to each other, the magnets will jump together.

UNLIKE POLES ATTRACT EACH OTHER

Turn the magnets so that LIKE POLES are next to each other. The magnets will jump apart. In fact, force is required to bring them close to each other, and they will jump apart if the force is removed.

LIKE POLES REPEL EACH OTHER

What happens to the lines of force when two magnets are placed in the attractive position (with unlike poles near each other)? This is illustrated in Figs. 7-3 and 7-4. Note the dense magnetic lines in the center between the two magnets.

Magnets placed in the opposing position (with LIKE poles together) and their magnetic fields are shown in Figs. 7-5 and 7-6. Note that the space between the magnets is vacant. Magnetic lines oppose each other and will not come close.

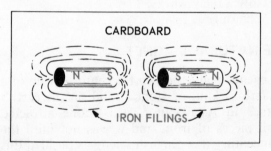

Fig. 7-5. When the magnets are placed in opposing position with like poles facing, no lines of force are visible between the poles.

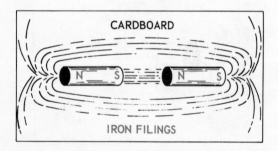

Fig. 7-3. When the magnets are placed in the attractive position with unlike poles facing, the lines of force combine.

Fig. 7-4. Photograph shows pattern of magnetic field when two magnets are in an attractive position.

Fig. 7-6. Photograph reveals effect of opposing magnetic fields on iron filings.

A MAGNET AND A NAIL

Why does a magnet pick up a nail? When the nail is close to the magnet, some of the magnetic lines will pass through the nail and cause the nail to become a MAGNET with poles necessary to cause attraction. Study Fig. 7-7. The nail, the magnetic fields and the poles produced will explain this action.

Interesting experiments and fascinating movements of magnets are produced by the device shown in Fig. 7-8. It is called "KRAZY LEG." The magnets may be arranged to suit your own fancy.

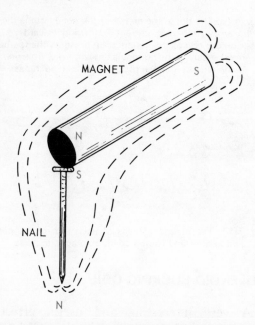

Fig. 7-7. The nail is attracted because the field of the magnet causes the nail to become magnetized.

ELECTRICITY AND MAGNETISM

In the early part of the 19th century, the Danish scientist, Hans Christian Oersted, was in his laboratory performing some simple experiments with electricity. A compass was nearby on the bench. Each time he turned on the electric circuit, the compass would swing on its pivot. The compass would also act up when the circuit was turned off. Thus, Mr. Oersted correctly assumed that his electric circuit must be affecting the compass. The long sought relationship between electricity and magnetism was discovered.

MAGNETIC FIELDS AND CONDUCTORS

Any wire or conductor carrying an electric current will create a magnetic field around the conductor and at right angles to the flow of current. This field of force forms concentric (having a common center) rings around the wire, like doughnuts strung on the wire. Fig. 7-9 illustrates these fields by dashed lines and arrows. The direction of the magnetic field depends on the direction of the current flow.

LEFT HAND RULE

The direction of the magnetic field caused by a current may be easily discovered by using the LEFT HAND RULE. Grasp a wire with your left hand with your thumb extended outward, pointing in the direction of current flow. Your fingers around the wire point in the direction of

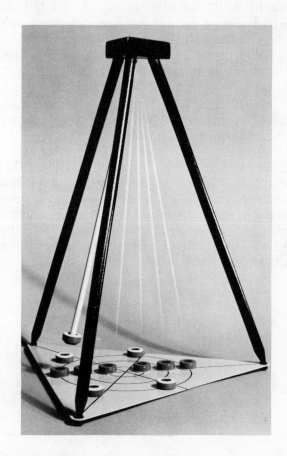

Fig. 7-8. This wild game can keep you interested for hours. Magnets on the base can be arranged to either attract or oppose the magnet on the swinging leg.
(Krazy Leg, Rathcon, Inc.)

Fig. 7-9. A current-carrying conductor has a magnetic field at right angles to current flow.

the field. The end view of a conductor and its magnetic fields may be observed in Fig. 7-10.

ONE FROM MANY FIELDS

Fig. 7-11 shows a wire wound into a coil. When current passes through the coil, magnetic fields are formed around the wire as indicated by small arrows. These join and reinforce each other, and the coil has a polarity like a MAGNET. This coil, since it has a hollow core, is called a SOLENOID. Which end is north or south? Again you may use your LEFT HAND RULE. Grasp the coil with your left hand with your fingers encircling the coil and pointing in the direction of current flow. Your extended thumb will point to the NORTH end of the coil.

EXPERIENCE 1. Prove that a solenoid does have magnetic polarity, and the polarity changes with the direction of current.

Connect a coil to a 6 volt supply, Fig. 7-12. Place a compass close to one end of the coil and observe the compass action. Reverse the connections to the power source. Does the compass change direction also? Explain this by using the LEFT HAND RULE. Does the compass point toward the north or south pole?

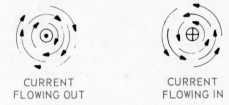

Fig. 7-10. The dot in the center of the conductor is the point of an arrow, meaning current is flowing toward you. The cross on the end of the conductor represents the feathers on an arrow, meaning the current is flowing away from you. Note the direction of the magnetic field in each case.

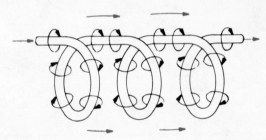

Fig. 7-11. The magnetic fields of each turn of the coil join together and form a coil with magnetic poles.

SOLENOID SUCKING COIL

A very interesting and useful effect of magnetism is demonstrated by the SOLENOID SUCKING COIL. This type of electrical device is widely used in home appliances and industrial

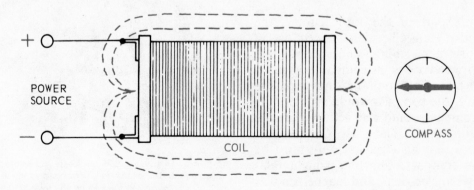

Fig. 7-12. When the coil is energized, the compass indicates the polarity of the magnetic field.

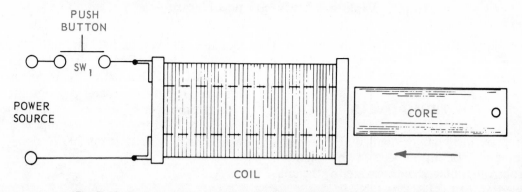

Fig. 7-13. Core moves inward by SUCKING ACTION when coil is energized.

machinery whenever it is necessary to operate some mechanical switch or valve by electric power.

EXPERIENCE 2. Set up a coil and attach it to the power source as shown in Fig. 7-13. Place an iron core part way into the coil. Turn on the power. What happens? The core is SUCKED into the coil and comes to rest at the center of the coil.

CONCLUSION: When the coil is energized, it produces a magnetic field. This magnetizes the iron core in the same manner as a magnet picks up the nail in Fig. 7-7. The core is attracted to the coil and they snap together. Once the core is in the center of the coil, the magnetic field is concentrated within the core. There is no room for further movement.

EXPERIENCE 3. Set up the same solenoid coil and attach an elastic band or spring to the end of core. See Fig. 7-14. Each time the circuit is closed by pushing button SW_1, the core is sucked into the center. When power is removed, the core returns to its original position.

OBSERVATIONS: In the solenoid, it seems that we have discovered a convenient method of converting electric power into straight line motion. The moving core could be used to turn a water faucet on or off. It could be used to strike a bell or chime, Fig. 7-15. It could be attached to a lever for shifting gears of a machine.

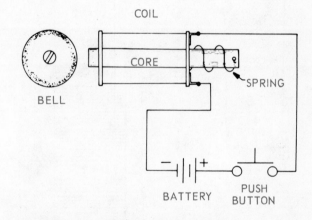

Fig. 7-15. The solenoid is used in the simple door chime. When the coil is energized, the core is sucked in and strikes the bell. The spring causes the core to return to its original position.

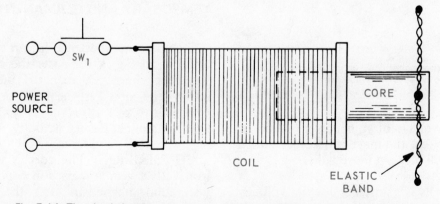

Fig. 7-14. The elastic band returns core to its original position when power is removed.

It might be used to activate a bolt lock on a door. Can you think of other ways to use a SOLENOID?

ELECTROMAGNETS

An improvement can be made on the strength of the magnetic field produced by the solenoid. If an iron core is placed in the center of the coil, it is called an ELECTROMAGNET.

EXPERIENCE 4. Assemble the coil and the core as instructed in Fig. 7-16. Connect it to your power source and adjust the voltage to about five volts. Check the coil with the compass to make sure a magnetic field does exist. Pick up several small nails or paper clips with the electromagnet. It works like a permanent magnet.

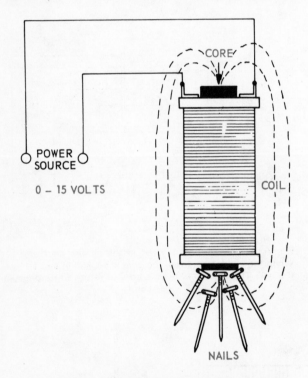

Fig. 7-16. The electromagnet picks up iron nails like the permanent magnet. An increase in voltage causes an increase in current and an increase in the strength of the magnet.

Increase the voltage of your power source to ten volts. Note that the magnet becomes much stronger and will pick up more nails.

CONCLUSIONS: An increase in voltage causes an increase in current and, therefore, an

increase in magnetic field strength. The strength of an ELECTROMAGNET is measured in AMPERE TURNS, which is abbreviated to IN. I is the current in amperes through the coil and N is the number of turns of wire in the coil. A coil with 200 turns and one ampere of current would have a strength of 200 IN:

$$1 \times 200 = 200 \text{ IN}$$

If current is increased to 2 amperes, the field strength would be 400 IN:

$$2 \times 200 = 400 \text{ IN}$$

MAGNETIC MATERIALS

Only a few materials will be attracted by a magnet. These include iron, nickel and cobalt. We have been using iron for a core in our experiences.

EXPERIENCE 5. Secure some samples of other kinds of materials like wood, plastic, zinc and copper. Try to pick up these materials with the electromagnet. The magnet does not work because these are not magnetic materials.

A boat builder went to a hardware store to purchase some brass screws which do not rust in water. The hardware dealer tried to sell the boat builder some brass plated screws. The boat builder has brought a small magnet on the trip. How did the boat builder discover that the screws were not brass? Hint: Brass is an alloy of copper and zinc.

TEMPORARY AND PERMANENT MAGNETS

Set up your experience for an electromagnet again as in Fig. 7-16. After the electromagnet has been energized, turn off the power and remove the core. Test the magnetism of the core by trying to pick up some nails. You will find that the magnet is very weak.

CONCLUSION: The core is made of soft iron with a very low carbon content and does not retain magnetism after the force which magnetized it has been removed.

EXPERIENCE 6. Repeat the activity, but this time use a steel core. Note that the steel will pick up nails after the magnetizing force has been removed.

CONCLUSION: Some materials, such as steel, will retain the magnetism after they have been magnetized.

Frequently, when working on electrical circuits with a steel screwdriver, the screwdriver will become magnetized. This can be a nuisance, since it attracts particles of iron and steel and loose nuts and bolts.

MIRACULOUS FERRITES

You should know something about some new materials made of iron oxide compounds. These are called FERRITES. These magnetic materials have almost, single-handedly, revolutionized the electronic industry. Without them, the development of television, computers and audio and video recorders would not have been possible.

A ferrite is an electrical insulator that can be magnetized in either a north-south or south-north polarity by applying a very small magnetizing force. Consequently, "ferrite memory cores" are widely used in the memory storage of the modern computer.

Ferrites are also used on audio and video recording tapes. The music or picture is converted into a varying electric current which magnetizes the tape in that particular pattern. For play-back, a sensing head interprets the pattern and produces an electrical potential corresponding to the recorded pattern. Presto, tape recorder.

THE MAGNETIC SWITCH

The RELAY or magnetic switch finds thousands of uses in electronic communications and industrial controls. It provides fast and positive switching for electrical circuits. It assures safety for workers because machines can be controlled from a remote location. Relays are economical to use. Only large and expensive wires are used near the machine, but the switching can be done by using low voltage and small wires.

A RELAY consists of four main parts, Fig. 7-17. They are the switching contacts, the armature, the electromagnet and a spring.

When a voltage is applied to the electromagnet, the magnetism attracts the armature and pulls it downward, causing the contact points to close. When the coil is demagnetized, the spring pulls the contact points apart and they return to their normal position.

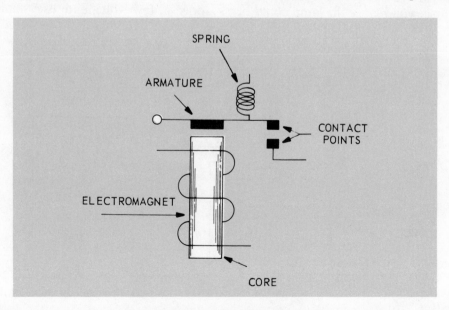

Fig. 7-17. The main parts of a RELAY.

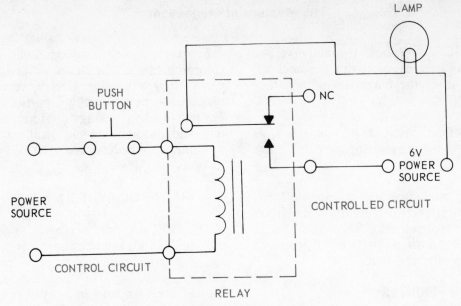

Fig. 7-18. The RELAY CIRCUIT: When the button is pushed, the coil is energized and the armature is attracted downward. The points are CLOSED and the lamp glows. NC means "no connection."

EXPERIENCE 7. The circuit for a relay is shown in Fig. 7-18. Note the schematic symbol used to represent a relay. Also note that this relay has another contact point. This relay can be used to turn a circuit ON or OFF depending on which contact points you use. The circuit in Fig. 7-18 is arranged to turn the lamp ON.

Now, rearrange the lamp connection so that the lamp is normally ON and will go OFF when you push the button.

How much voltage is required to operate the relay? Of course, you realize the lamp could be a motor or some other high voltage or high current device.

EXPERIENCE 8. This time, we will arrange the circuit using two lamps. L_1 will be ON and L_2 will be OFF. Study Fig. 7-19. When the button is pushed, L_1 will go out and L_2 will turn ON. Is there any electrical connection between the INPUT or controlling circuit and the OUTPUT or controlled circuit? This is one of the major advantages of relay circuits.

CONCLUSION: The input circuit could be only a few volts and use small, inexpensive

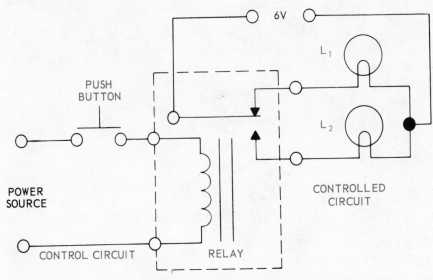

Fig. 7-19. Relay circuit L_1 is ON and L_2 is OFF. When button is pushed, L_1 goes OFF and L_2 is turned ON.

78

wires, while the output or controlled device could require large wires, heavy currents and several hundred dangerous volts. Of course, the push button switch could be miles away and the operator would be safe from any dangers.

THE REED RELAY

The reed relay is another similar application of magnetism. In Fig. 7-20, note two magnetically sensitive switch contacts are enclosed in a glass tube. If a permanent magnet is brought close to the glass tube, it will cause the switch contacts to close. The reed relay can also be operated by an electromagnet. The operating coil is placed around the reed relay.

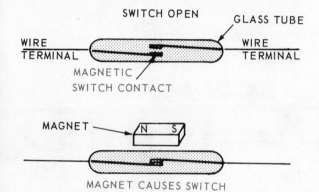

Fig. 7-20. A sketch of a reed relay. The magnet causes the reed contacts to close.

A BUZZER

A buzzer or doorbell uses electrical principles similar to the relay. The difference is in how the connections are made.

EXPERIENCE 9. A circuit for a BUZZER is drawn schematically in Fig. 7-21. An interesting experience for you would be to build this buzzer and observe its action.

CONCLUSIONS: When the coil is energized by pushing the button, the armature is attracted downward. This causes the points to open which, in turn, "opens" the circuit. The coil magnetism drops to zero and the contact points are allowed to close. This again energizes the circuit and the points open. This action continues as long as a voltage is applied to the buzzer (so called, because it buzzes).

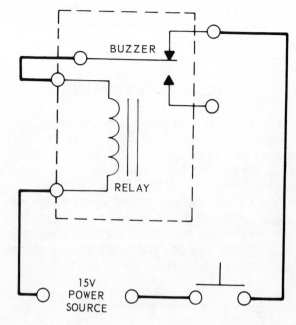

Fig. 7-21. The circuit for a BUZZER.

CHOPPER

Consider that the current flowing through the buzzer circuit is not a continuous current. In fact, when the contact points open, the current drops to zero. When the points are closed, the current is maximum. So the current is rising and falling as the armature vibrates. Fig. 7-22 shows a simplified graph of the current in the circuit. It looks like a SQUARE WAVE.

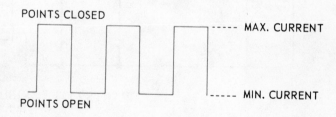

Fig. 7-22. This square wave shows the action of the buzzer-chopper in a circuit.

Up to this point in your studies, only dc or DIRECT CURRENT has been discussed. Direct current remains at a CONSTANT VALUE depending on the resistance of a circuit. The Chopper produces a PULSATING direct current. You will recall, from your studies, that a steady direct current produces a magnetic field. A pulsating direct current produces a PULSATING MAGNETIC FIELD. This principle is in extensive use in power supplies, automotive ignition systems and electronic measuring circuits.

BURGLAR ALARM PROJECT

Home security systems are an important segment of the electronics industry today. This project uses an inexpensive, yet effective, circuit to detect intruders. See Fig. 7-23.

The basic circuit uses a light source and a cadmium sulfide (CDS) photocell in an alarm circuit. Refer to Fig. 7-24 for the schematic diagram. Be sure that the light source is placed across from the alarm unit so that the intruder

Fig. 7-23. Burglar Alarm.

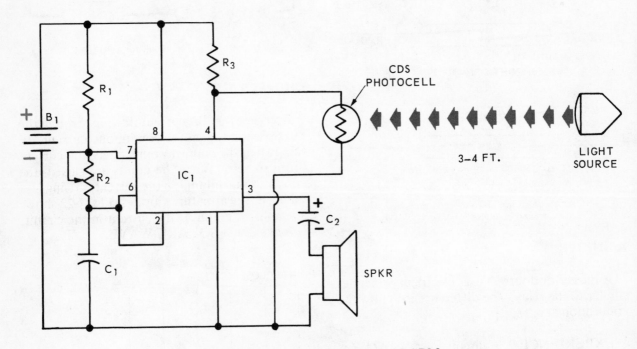

PARTS LIST FOR BURGLAR ALARM

R_1 — 1 KΩ, 1/2W resistor
R_2 — 500 KΩ, 2W potentiometer
R_3 — 100 KΩ, 1/2W resistor
C_1 — capacitor, .01 μF, 50 WVdc
C_2 — electrolytic capacitor, 4.7 μF, 16 WVdc
IC_1 — integrated circuit, NE 555 or LM 555
SPKR — 2 in. 8 Ω speaker

CDS — cadmium sulfide photocell, 200 MW, resistance (dark) 5 meg. Ω, resistance (light) 100 Ω
B_1 — power supply or 9V battery
Light source — incandescent lamp (117V) with lens in metal box
Misc. — P.C. board with materials, wire, solder, decals

Fig. 7-24. Schematic and parts list for Burglar Alarm.
(Reprinted from POPULAR ELECTRONICS MAGAZINE, Ziff-Davis Publishing Co.)

will "break" the light beam. The potentiometer is used as a tone control. A 9 volt battery or power supply can be used to operate the device.

FORWARD STEPS IN UNDERSTANDING ELECTRICITY-ELECTRONICS

1. A magnet has a magnetic field and a NORTH and a SOUTH pole.
2. Like magnetic poles repel each other.
3. Unlike magnetic poles attract each other.
4. A magnetic field exists around a current-carrying conductor. The field is at right angles to the flow of current.
5. The direction of the magnetic field around a current-carrying conductor may be found by using the Left Hand Rule.
6. A coil without a core is called a solenoid.
7. A coil with a core is called an electromagnet.
8. The strength of an electromagnet is measured in ampere-turns (IN).
9. A permanent magnet is made by placing a material which retains its magnetism in a strong electromagnetic field.
10. A relay is a magnetic switch.
11. Ferrites are iron compounds. They can be easily magnetized and demagnetized.
12. Direct current may be steady or pulsating.
13. A buzzer operates on magnetic action and "make and break" contact points.

TEST YOUR KNOWLEDGE - UNIT 7

1. Secure a 6 volt lamp, a relay, a switch and about ten feet of double conductor wire. Use an available power source. Build a remote control circuit to turn a lamp ON by a relay, using a switch at a distance of ten feet.
2. Secure a permanent magnet and some nails. Prove that a magnet will attract nails. Place a piece of window glass between the magnet and the nails. Does the glass affect the magnetism? Place a piece of copper between the magnet and the nails. Does copper allow the magnetic field to pass?
3. Connect an electromagnet to a power source with an ammeter in series. Adjust voltage to five volts. What is the current flow? Secure another iron core. Increase voltage until the

electromagnet will pick up and hold the second core. Decrease the voltage slowly watching the current meter. At what current is the magnetic field sufficiently weak to drop the second core? What does this experience prove to you?

4. Wire up a doorbell or buzzer circuit with push buttons at the front and rear door. Use your relay as a buzzer. Use two push buttons and two pieces of double conductor wire. Either button must ring the buzzer.
5. Build an overcurrent indicator. Connect circuit by consulting schematic in Fig. 7-25. Note that the relay is in series between the power source and load. Set voltage at 15 volts. Adjust load resistance for "closing current" of relay. At this point L_1 will glow. Current = _____.

   Increase load R and reduce current until relay opens. Minimum holding current = _____.

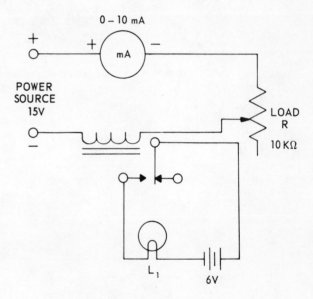

Fig. 7-25. Circuit for problem 5.

6. Construct the relay buzzer circuit shown in Fig. 7-21. Ask your instructor to connect an oscilloscope across the coil. Draw a picture of the wave form.
7. Make a list of five ways that a solenoid might be used as an electrical-mechanical control.

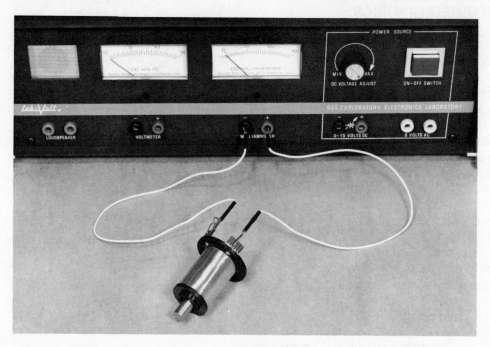

Fig. 8-1. A laboratory setup for performing the experience of generating electricity.

Unit 8

DIRECT AND ALTERNATING CURRENT

The two basic types of currents in the field of electricity are direct current (dc) and alternating current (ac).

This Unit will cover dc, ac and the following major topics:

1. How Faraday's discovery helped advance the field of electricity.
2. What alternating current is.
3. How a simple generator (alternator) produces ac.
4. What direct current is.
5. How a simple generator produces dc.
6. How a transformer works.
7. What a power supply is, and how it operates.
8. What the process of rectification is, and how it is accomplished through diodes.
9. How a filter operates.
10. How a loud speaker operates.

GENERATING ELECTRICITY

In our study of Sources of Electrical Energy, the use of a magnetic field and a moving conductor was purposely delayed. This important source accounts for most of the electrical power now used in homes and industry. The development and growth of our industrial and military giants can be attributed to the discovery of MAGNETIC INDUCTION by Michael Faraday early in the 19th century.

Based on the research and discoveries of Faraday, the electric dynamo was developed. Faraday, in fact, is known as the "father of the dynamo."

This Unit can be one of your more important studies of electricity. To assure your understanding, perform the following experiences and "rediscover" the principles which made Faraday famous:

EXPERIENCE 1. A coil is connected directly to a milliammeter as in Figs. 8-1 and 8-2. The milliammeter used must have a pointer which can move either to the right or left of zero. Take a magnet and hold it above the hollow core of the coil. Watch the meter as you quickly push the magnet into the coil. Which way does the meter move? To the right or to the left? Now quickly withdraw the magnet from the core. Which way does the meter move? To the right or to the left? Does the meter indicate any current in either direction when the magnet is not moving?

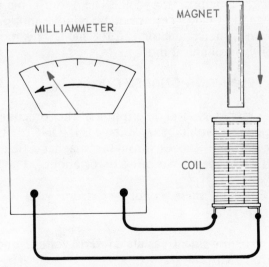

Fig. 8-2. Coil is attached to milliammeter to demonstrate the generation of electricity.

CONCLUSIONS: We have a coil of conductors and a magnetic field close to each other. Nothing happens unless there is movement of the magnet. Movement of the magnet downward into the coil produces a current, as read on the meter, in ONE DIRECTION. Pulling the magnet out of the coil produces a current in the OPPOSITE DIRECTION. You have made a simple GENERATOR.

EXPERIENCE 2. Repeat Experience 1 with these changes. Hold the magnet in a fixed position and MOVE THE COIL DOWN OVER THE MAGNET. Does the meter pointer move to the right or the left? Pull the coil from the magnet. Does the meter pointer move to the right or left? Does the meter move at all when there is no movement of the coil?

CONCLUSIONS: Electricity can be generated by movement of either the magnetic field or by moving the coil. THE DIRECTION OF THE CURRENT DEPENDS UPON THE DIRECTION OF MOVEMENT.

FARADAY'S DISCOVERY

Faraday's discovery of magnetic induction showed a definite relationship between a magnetic field and electricity. This, in turn, led to the development of the generator.

A GENERATOR produces electrical energy from mechanical energy. There must be a conductor, such as the coil. There must be a magnetic field. There must be relative motion between the conductor and the field. This motion is mechanical motion.

ALTERNATING CURRENT (ac)

A more technical description of the foregoing experience would be: A voltage is INDUCED in the coil as it moves through a magnetic field. INDUCE means "to bring on or about."

Consider these conditions from your experience:

1. No movement results in zero voltage and zero current.
2. Downward movement results in voltage of one polarity and current in one direction.

3. Upward movement results in voltage of opposite polarity and current in opposite direction.

EXPERIENCE 3. The voltage induced in Experiences 1 and 2 is illustrated in Fig. 8-3. The curve represents the magnitude or greatness of the voltage and the polarity of the voltage during one cycle of events. A CYCLE is an interval of time in which one round of events is completed. This cycle of events is: Starting with the magnet out of the coil, pushing it into the coil, withdrawing the magnet from the coil and back to the starting point.

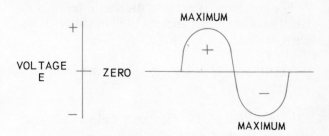

Fig. 8-3. This illustration shows the induced voltage and polarity during one cycle.

CONCLUSION: During the cycle, the polarity of the voltage and the direction of current flow will change. If you continued to move the magnet in and out, an ALTERNATING VOLTAGE and CURRENT would be produced. Fig. 8-4 shows the resulting voltage induced during four cycles.

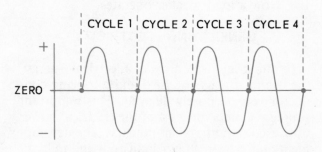

Fig. 8-4. The induced voltage and polarity for four complete cycles.

FREQUENCY

FREQUENCY simply means how often the cycle of events happen. Assuming continuous movement between the magnetic field and the coil, if the one cycle was completed in every second, the frequency would be:

ONE CYCLE PER SECOND OR ONE HERTZ

HERTZ is the unit of measurement for frequency. It means "cycles per second." The electricity we use at home alternates 60 cycles per second or 60 Hz (note abbreviation). Sound waves from a Hi-Fi STEREO will have frequencies from 20 to 20,000 Hz or 20 KHz (note use of prefix kilo).

Radio waves used in the AM broadcasting band have frequencies between 550 KHz and 1600 KHz. Look at the dial on the transistor radio and you will see these numbers. On a FM radio, note that the dial reads from 88 MHz to 108 MHz (M means mega or one million).

Television channels are six megahertz wide. Thus, a TV set tuned to Channel 4 uses the frequencies between 66 and 72 MHz. Channel 12 uses 204 to 210 MHz.

SIMPLE GENERATOR

From the practical point of view, pushing a magnet in and out of a coil would be a bit

tiresome. It is better to use a fixed magnetic field and rotate a coil in the field as described in Fig. 8-5. Only a single turn of wire is illustrated for ease in understanding. The ends of the coil are brought out to copper SLIP RINGS. BRUSHES rubbing on these slip rings bring the generated electricity to an external circuit and load.

Study Fig. 8-5. Coil side A moves downward and side B moves upward as the coil rotates through its first one-half revolution. Current flows as indicated by arrows. During the second half revolution, side B moves downward and side A moves upward through the magnetic field. As a result, the induced voltage and current reverses direction. The output per revolution or per cycle is similar to Fig. 8-3. This machine is called an ALTERNATOR. It produces an alternating current.

ALTERNATOR

From the electrical point of view, it would be better not to run the generated currents through sliding contacts like the brush and slip ring. Therefore, the practical alternator generally has the coils in a fixed position and rotates the magnetic field.

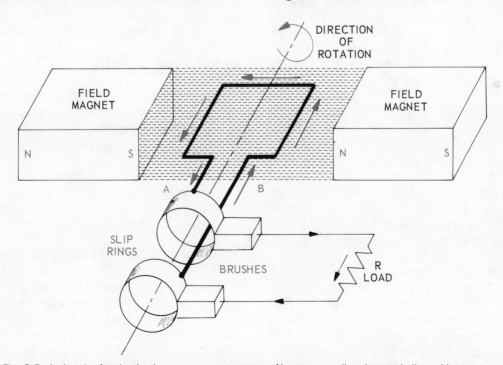

Fig. 8-5. A sketch of a simple alternator or ac generator. Note current directions as indicated by arrows.

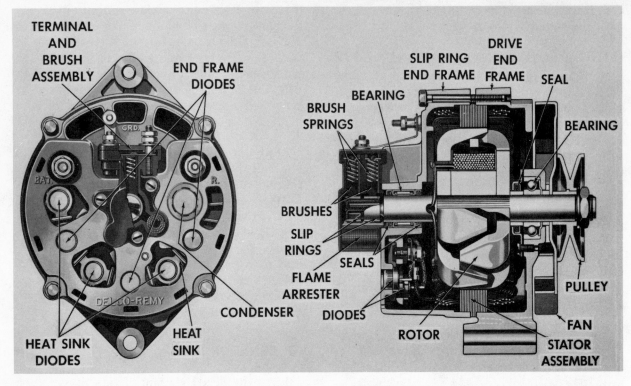

Fig. 8-6. Cutaway view of modern automotive ALTERNATOR. Study the various parts described in the text. The purpose of the diodes will be discussed later under rectification.   (Delco-Remy)

The fixed coils are called the STATOR and the windings of the revolving magnetic field are the ROTOR windings. The ROTOR does have slip rings, but they are connected to a direct current power source and only serve to excite the magnetic field. An alternator from a modern automobile is shown in Fig. 8-6.

DC GENERATOR

All generators produce an alternating current within their rotating armature. In order to realize a direct current output, the ac output must be RECTIFIED. This means it must be changed to dc. The dc generator uses a mechanical switch called a COMMUTATOR for this purpose. Study the diagram in Fig. 8-7.

Follow the action. Brush A is in contact with commutator section A. Also, brush B and section B are together. Current through the first half revolution flows as indicated by arrows. When the current in the armature reverses during the second half revolution, the commutator has also revolved and now section B is with

brush A and section A is with brush B. The current, therefore, continues to flow in one direction even though it has alternated in the armature. The output is a pulsating direct current.

GENERATOR OUTPUT

To improve upon the simple generator and increase its output, several things are done.
1. A stronger magnetic field is produced by using electromagnets. This increases the output.
2. Several coils of many turns of wire are used in the armature, together with the required commutator sections. This increases the output and makes it less pulsating.

GENERATOR OUTPUT REGULATION

The speed of rotation of the generator or alternator does affect the output. However, it is not always convenient to change the speed. The output of these machines, both current and voltage, is regulated by varying the strength of the magnetic field. Mechanical relays and tran-

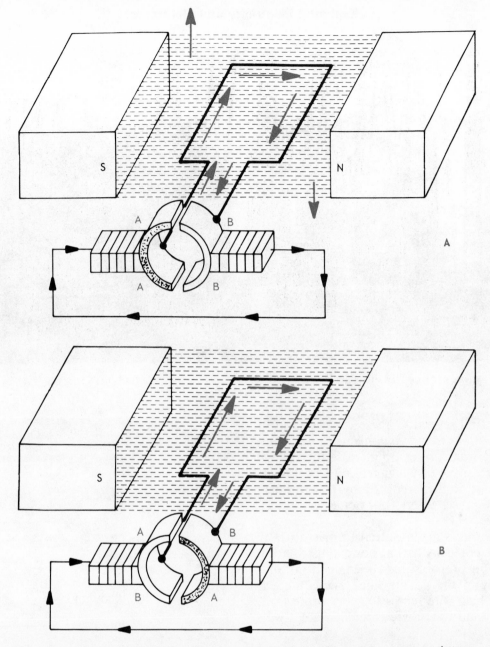

Fig. 8-7. A sketch of a dc generator. The commutator switches the generator output so that current in the output circuit flows only in one direction.

sistors are used to electronically switch more or less resistance in the field circuit and vary the field current. This varies the field strength and the generator output. Automobiles have such current-voltage regulators. An example is shown in Fig. 8-8.

TRANSFORMERS

The TRANSFORMER is an electrical component which transfers electrical energy from one circuit to another by using magnetic fields.

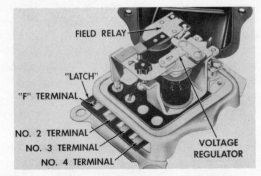

Fig. 8-8. A typical relay type voltage regulator used to control generator output.   (Delco-Remy)

Fig. 8-9. This power distribution center has a transformer bank of four General Electric single-phase auto transformers. (Pacific Gas and Electric Co.)

You will find several of these devices in most any kind of electronic equipment. Power companies use transformers to increase the voltage for their cross-country high tension power lines. The voltage is then reduced by transformers before the power enters your home.

Fig. 8-9 shows the large transformers used by a power company at a power distribution center. You probably have seen such installations around the countryside. These are enclosed by high wire fences for your protection. People should treat these forbidden areas with respect.

Now, to discover how a transformer works. First, remember Faraday's discovery: To induce voltage, there must be a coil, a magnetic field and RELATIVE MOTION between the coil and the field.

Study Fig. 8-10 in which a coil is attached to an ALTERNATING CURRENT source. We already know that an increased voltage will increase the magnetic field around the coil. At point A of the input voltage wave, the voltage is zero and there is no magnetic field. As voltage increases to point B, the magnetic field also moves outward or EXPANDS to its maximum strength at a definite polarity.

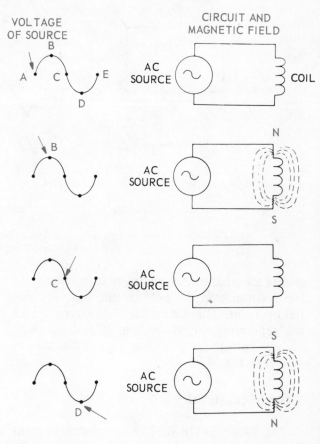

Fig. 8-10. These circuits illustrate the changing magnetic field of a coil as an ac voltage is applied to it.

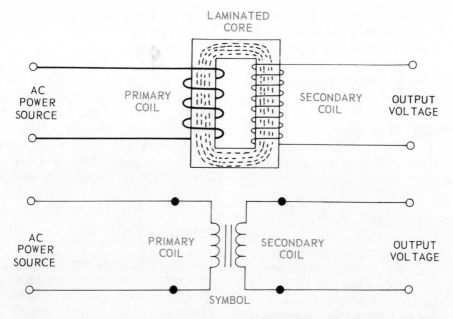

Fig. 8-11. Sketch and symbol for a transformer showing primary and secondary windings.

As voltage decreases to point C and zero, the magnetic field moves inward or collapses to zero. At point D, the voltage again builds up to maximum in the OPPOSITE DIRECTION and the magnetic field again EXPANDS outward to maximum strength, but with OPPOSITE POLARITY. At point E, voltage has returned to zero and field has collapsed to zero.

A continuing ac voltage would produce a continuing expanding and collapsing magnetic field. Therefore, we have the MAGNETIC FIELD required in Faraday's discovery AND we have MOVEMENT OF THE MAGNETIC FIELD (outward and inward) produced by using an ac voltage. All that remains are the necessary conductors.

To meet this need, a second coil of wire is placed close to the first coil. It usually is wound directly over the first coil. Both coils are wound on a common, laminated iron core. This means the core, rather than being solid, is made up of thin layers of core material. Plywood is a good example of laminating. Lamination reduces losses in the transformer.

The symbol for the transformer with its two coils and laminated core is given in Fig. 8-11. Typical power transformers used in electronics are illustrated in Fig. 8-12. THE INPUT COIL IS THE PRIMARY WINDING AND THE OUTPUT COIL IS THE SECONDARY WINDING.

ACTION IN THE TRANSFORMER

When the primary coil is connected to an ac source of power, THE EXPANDING AND COLLAPSING MAGNETIC FIELD CUTS

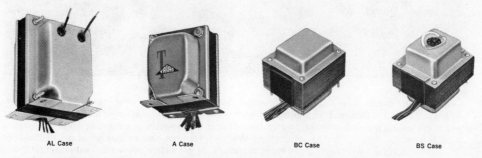

AL Case   A Case   BC Case   BS Case

Fig. 8-12. Typical types of power transformers used in electronics.   (Triad)

ACROSS THE SECONDARY WINDING AND INDUCES A VOLTAGE IN THE SECONDARY. The output voltage depends on the number of turns of wire in the primary in respect to the number of turns in the secondary. This is called the TURNS RATIO.

Neglecting losses, if both primary and secondary have the same number of windings, the turns ratio is 1 to 1. The output voltage is the same as the input voltage.

STEP-UP TRANSFORMER

By increasing the turns of the secondary, a higher output voltage may be produced. For example: Assume the primary has 500 turns and the secondary 2000 turns. The ratio is now 1 to 4. An input voltage of 100V ac would produce an output of 400V ac. This is a STEP-UP transformer, which is used to change a voltage to a higher voltage to meet the needs of certain equipment.

STEP-DOWN TRANSFORMER

As you might expect, if there are less windings on the secondary than on the primary, the output voltage is lower than the input voltage. For example: Again assume that the primary has 500 turns, but the secondary has only 100 turns. Now the ratio is 5 to 1. An input voltage of 100V ac would produce 20V ac in the secondary or output. These STEP-DOWN transformers also are extensively used in electronic circuits.

TURNS RATIO

A simple formula may be set up by which you can find the turns ratio (number of turns of wire in primary in respect to number of turns in secondary) or the output voltage:

$$\frac{E_{in}}{E_{out}} = \frac{N_{primary}}{N_{secondary}}$$

Circuits for air safety. Every second around the clock, two airplanes either take off or land somewhere in the United States. This photograph shows the work station at an air traffic control center, where planes are tracked from the time of takeoff until they reach a point within 15 or 20 miles of their destination, where airport control takes over. All this time, the various pilots get instructions from controllers who keep them in flight paths that are safely away from other aircraft. (Bell System)

In this formula:

E_{in} = input voltage

E_{out} = output voltage

$N_{primary}$ = number of turns in primary coil of transformer

$N_{secondary}$ = number of turns in secondary coil of transformer

Work out the previous examples using this formula.

For the step-up transformer:

$$\frac{100}{E_{out}} = \frac{500}{2000} \qquad E_{out} = 400V \ ac$$

For the step-down transformer:

$$\frac{100}{E_{out}} = \frac{500}{100} \qquad E_{out} = 20V \ ac$$

The symbol for a power transformer which has both step-up and step-down windings is shown in Fig. 8-13. These transformers are designed to meet the requirements of a certain electronic circuit.

Transformers are also used to transfer power from one stage of a circuit to another. This is called TRANSFORMER COUPLING.

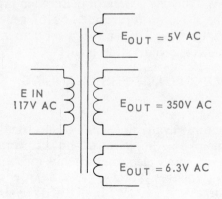

Fig. 8-13. The symbol of a transformer with both step-up and step-down windings.

POWER TRANSFORMERS

Back in Unit 4, the very important subject of power as discussed. You should remember that:

$$\text{POWER (in watts)} = \text{VOLTAGE} \times \text{CURRENT}$$

With this in mind, the circuit for a step-up transformer will be found in Fig. 8-14. In this case a load of 500 ohms is attached across the

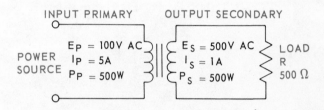

Fig. 8-14. The power output must equal the power input of a transformer. Losses are not considered.

secondary. A current will flow. Assuming 100 volts ac input and a 1 to 5 step-up transformer, the secondary voltage is 500 volts. The current in the secondary equals:

$$I_{secondary} = \frac{E_{secondary}}{R \ ohms}$$

or

$$I = \frac{500V}{500 \ \Omega} = 1 \ amp$$

Power used by secondary therefore is:

$$I_{secondary} \times E_{secondary} = P_{secondary}$$

or

$$I \ A \times 500V = 500 \ watts$$

Of course, if the secondary circuit is using 500 watts of power, then the PRIMARY must supply 500 watts of power. A TRANSFORMER IS NOT A SOURCE OF POWER. Then 500 watts of power is drawn from the source. But the voltage of the primary is only 100 volts. How much current must flow in the primary to realize 500 watts?

$$I_{primary} \times E_{primary} = P_{primary}$$

or

$$I_{primary} \times 100V = 500$$

and

$$I_{primary} = 5 \text{ amps}$$

To prove this:

POWER INPUT TO TRANSFORMER
= POWER OUTPUT

Then:

$$I_{primary} \times E_{primary}$$
$$= I_{secondary} \times E_{secondary}$$

or

$$5A \times 100V = 1A \times 500V$$

and

$$500 \text{ watts} = 500 \text{ watts}$$

POWER TRANSMISSION LINES

Why does the power company use transformers to increase the voltage of their cross-country high tension lines? To answer this, we must look into the subject of POWER LOSS: When current flows through a conductor, it must overcome the resistance of the conductor. This produces HEAT, which serves no useful purpose. It is lost to the surrounding air. It is "good business" to keep LOSSES AT A MINIMUM. If current is causing the loss, then current must be kept at a very low value.

For example, assume that a small factory needs 10,000 watts of power at 117 volts. The current required would be:

$$I \times 117 = 10,000 \qquad \text{or} \qquad I = 85.5 \text{ amps}$$

This is a lot of current. The wires would have to be large and the losses would be great to bring this current from a faraway generator. To avoid this, the voltage is transformed (stepped up) to

a very high voltage, sent across country and then transformed (stepped down) to the lower required voltage.

Continuing the example, assume the voltage is increased to 200,000 volts. With this voltage, the current for the needed power is:

$$I \times 200,000 = 10,000W \qquad I = .05A$$

Do we have the same power? Compute the formula both ways:

$$85.5A \times 117V = 10,000 \text{ watts}$$
or
$$.05A \times 200,000V = 10,000 \text{ watts}$$

By reducing the current in the high tension lines, the power loss is kept at a minimum. The complete system is drawn in Fig. 8-15. Note the use of transformers to increase voltage and decrease current.

At the generator $I \times E = 1200W$ or
$.1A \times 12,000V = 1200W$
Cross-Country $I \times E = 1200W$ or
$.02A \times 60,000V = 1200W$
City Lines $I \times E = 1200W$ or
$.1A \times 12,000V = 1200W$
Home $I \times E = 1200W$ or
$10.25A \times 117V = 1200W$

Fig. 8-16 illustrates the use of power lines to carry electrical power cross-country at a very high voltage. Fig. 8-17 shows a hydro-electric generating station at Shasta Dam in California. The wind turbine generator pictured in Fig. 8-18 may be a major future method of generating electricity. Fig. 8-19 shows a modern nuclear steam-electric generating plant.

POWER SUPPLIES FOR CHANGING AC TO DC

The most available source of electricity is ac provided by the electric power company. However, many electronic circuits must operate from dc voltages. In these cases, a power supply may be used. A POWER SUPPLY is an electronic circuit that provides the necessary voltages and currents to operate other electronic circuits.

Direct and Alternating Current

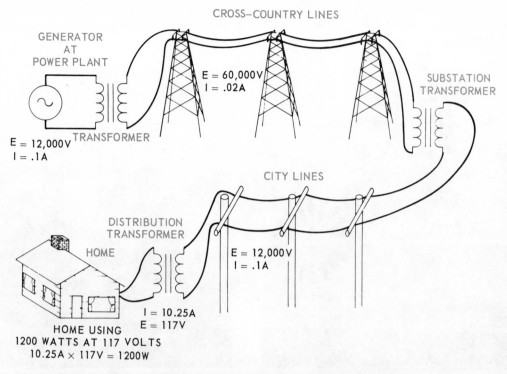

CROSS-COUNTRY LINES

GENERATOR AT POWER PLANT

E = 60,000V
I = .02A

SUBSTATION TRANSFORMER

E = 12,000V
I = .1A

TRANSFORMER

CITY LINES

DISTRIBUTION TRANSFORMER

HOME

E = 12,000V
I = .1A

I = 10.25A
E = 117V

HOME USING
1200 WATTS AT 117 VOLTS
10.25A × 117V = 1200W

Fig. 8-15. This system shows how voltage is stepped up so that current can be minimum. This reduces power loss in the transmission lines. Voltage and current values selected to illustrate text example. High tension lines may carry 500,000 volts.

Fig. 8-16. Power lines carry electricity from a nuclear generating station. (Carolina Power & Light Co.)

Fig. 8-17. Hydro-electric generating station at Shasta Dam. (Bureau of Reclamation)

Fig. 8-18. A large, 200 kilowatt wind turbine generator used in Clayton, New Mexico.   (U.S. Department of Energy)

Fig. 8-19. A nuclear generating plant in Italy. (Office of European Atomic Energy Community)

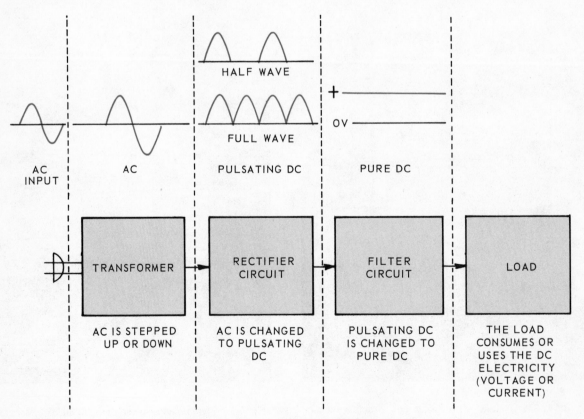

Fig. 8-20. Power supply block diagram.

Devices such as radios, television sets, citizens band radios and other electronic equipment have power supplies built into their circuitry. A typical block diagram for a power supply is shown in Fig. 8-20.

Some power supplies are constructed separate from the circuits they operate. Many laboratory supplies are of this type. See Fig. 8-21.

Fig. 8-21. Typical laboratory power supply.

RECTIFICATION

Power is supplied to our homes by local power companies at 117 and 235 volts at a frequency of 60 Hertz. But much of the electronic equipment like radios and television sets require direct current rather than alternating current. Also voltages other than 117 volts are needed. This introduces you to the means of changing ac to dc. This process is called RECTIFICATION.

Rectification is accomplished in electronics through a device called a RECTIFIER or DIODE. Refer to Fig. 8-22 to see ac waves being converted to dc through a rectifier.

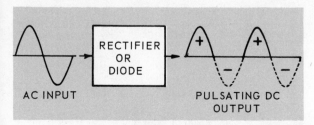

Fig. 8-22. Rectifier action changes ac to dc.

DIODE RECTIFIERS, A ONE-WAY STREET

Like a one-way sign, a DIODE RECTIFIER permits the flow of traffic in one direction only. A DIODE is a one direction device which permits current to flow easily in one direction, but offers an extremely high resistance to a current trying to flow in the opposite direction.

The most common type of diode used today is the solid state unit consisting of two types of silicon crystals. (Crystal diodes will be studied in Unit 10.) The symbol for the diode and a basic circuit which shows its operation are shown in Fig. 8-23.

Diodes and rectifiers usually are rated according to their current and voltage capabilities, as well as their peak reverse voltage (PRV).

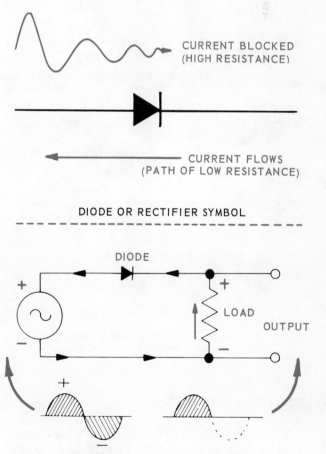

Fig. 8-23. The diode permits current to flow in one direction only. When voltage of the source changes polarity, the current drops to zero.

Fig. 8-24 shows typical diodes used in a rectifier circuit. These diodes are made in many types, sizes and current ratings. The output voltage appears across the diode load, and we know that voltage will be:

$$I_{circuit} \times R_{load} = E_{output}$$

Assuming an ac source with a 10 volt effective voltage and a load resistance of 1000 Ω, the current would be .01 amperes. The output voltage would be:

$$I \times R = E \quad \text{or} \quad .01 \times 1000 = 10 \text{ volts}$$

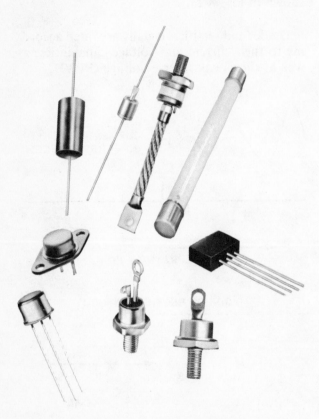

Fig. 8-24. Assorted type of diode rectifiers and silicon controlled rectifiers. (International Rectifier Corp.)

RECTIFIER CIRCUITS

SINGLE DIODE RECTIFIER. The simplest type of rectifier circuit uses a single diode. This circuit blocks the current flow in one direction and passes it in the opposite direction. The output wave is half-wave dc. Fig. 8-25 shows a SINGLE DIODE RECTIFIER circuit.

EXPERIENCE 1. Secure a diode and 1000 Ω resistor, connect them to a 6.3 volt ac power source. See Fig. 8-25. Connect an oscilloscope first to the ac source. Next, adjust the oscilloscope until four full ac waves appear on the CRT. See page 185. Then, switch the oscilloscope controls to the 10 volt range. The pattern should appear as in Fig. 8-26.

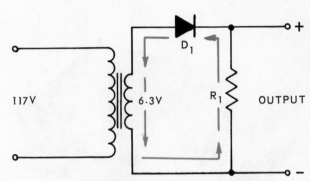

Fig. 8-25. Single diode rectifier circuit.

EXPERIENCE 2. Leaving the oscilloscope adjustments as before, change the test leads to the output of the diode rectifier circuit. Compare to the pattern in Fig. 8-26.

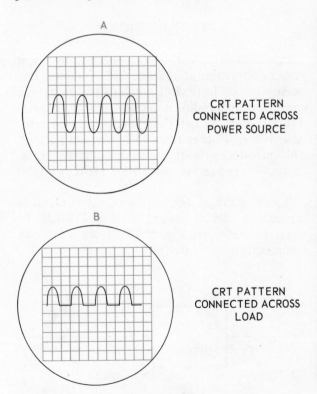

Fig. 8-26. A comparison of input and output wave forms of the diode rectifier circuit.

CONCLUSIONS: The output voltage no longer goes in a NEGATIVE direction. Therefore, the output is a PULSATING DIRECT CURRENT and not an alternating current. The PEAK output voltage is approximately the same as the PEAK input voltage. Since one-half of the ac input wave is cut off and contributes nothing to the output, the circuit is called a HALF-WAVE RECTIFIER.

FULL-WAVE RECTIFIER. In order to secure full-wave rectification, two diodes must be used. In full-wave rectification, both half cycles of the ac input will produce a useful output. However, the voltage output will be only one-half as much unless a more elaborate bridge rectifier with four diodes is used.

EXPERIENCE 3. Construct a full-wave rectifier by following the schematic diagram in Fig.

8-27. The transformer is a step-down to 12.6 volts for the secondary. The secondary also has a center tap. In Fig. 8-27, the top of the secondary is positive and diode D_1 conducts, producing an output voltage across the load.

In Fig. 8-28, the bottom of the secondary is positive and D_2 conducts and produces output voltage across R. Therefore, there is output during both halves of the ac input wave. The voltage output is approximately 6.3 volts since it is taken from only one-half of the secondary winding.

Connect the oscilloscope across the secondary winding and adjust the controls to display four complete wave forms. See page 185. Now, connect the oscilloscope across R and adjust to display eight half waves. Waves should appear as in Fig. 8-29.

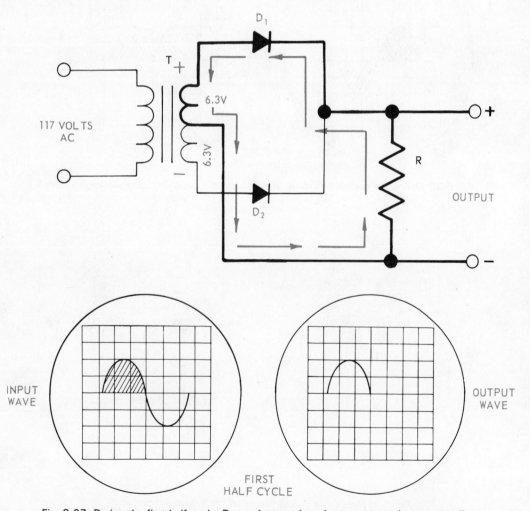

Fig. 8-27. During the first half cycle, D_1 conducts and produces output voltage across R.

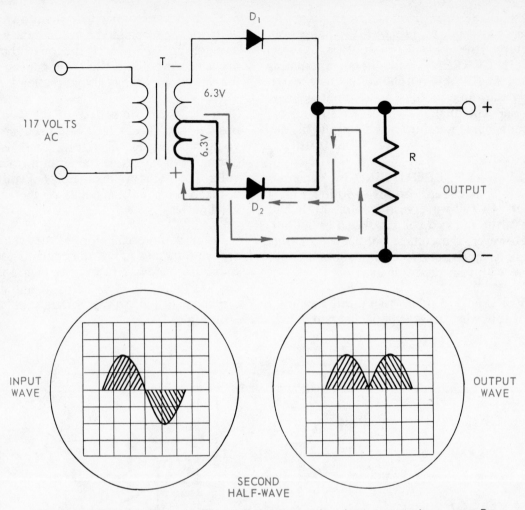

Fig. 8-28. During the second half cycle, D₂ conducts and produces output voltage across R.

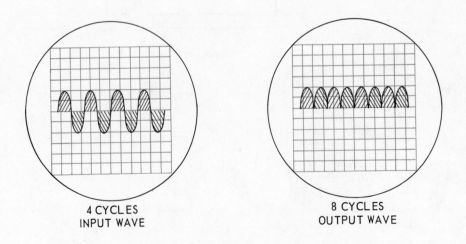

Fig. 8-29. Input and output waves of the full-wave rectifier circuit. The output is two times the frequency of the input.

CONCLUSION: The output wave of the half-wave rectifier has twice the frequency of the input wave, and it is a pulsating direct current.

BRIDGE RECTIFIER

The bridge rectifier provides full-wave rectification without the use of a center-tapped transformer. One of the characteristics of a bridge rectifier is that it uses four diodes. These may be four separate diodes or they may be molded or encapsulated into one assembly. Study diagrams A and B in Fig. 8-30 to observe the current flow in a bridge rectifier circuit.

FILTERING

Observe the roughness of both the half-wave and full-wave rectifier waves. Even though they are dc, they nevertheless vary between zero and peak value. A constant value of dc voltage is desired.

Fig. 8-31 shows a complete power supply, consisting of a transformer and a rectifier followed by a filter. The action of the capacitors is important. At peak pulses, the capacitor charges to peak voltage.

When the pulse from the rectifier drops to zero, the capacitors discharge into the load. The

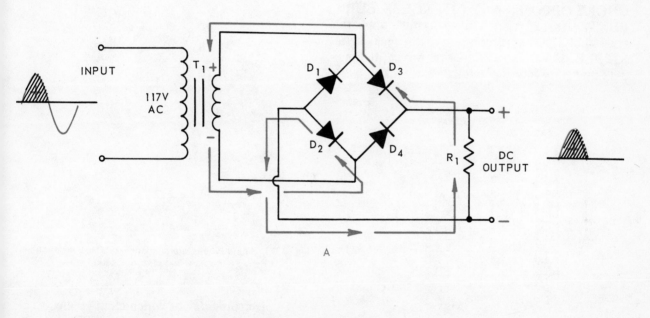

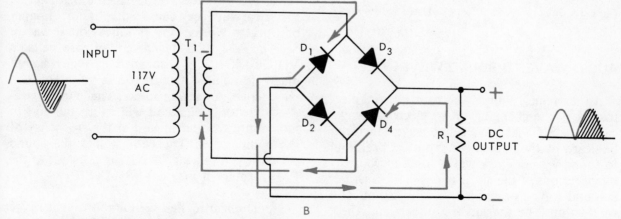

Fig. 8-30. Full-wave rectification: A—Bridge rectifier current path. B—Opposite bridge rectifier current path.

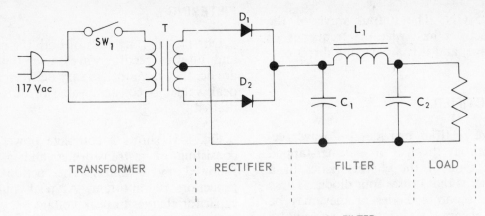

Fig. 8-31. The complete power supply with FILTER.

filtering action tends to hold the output across the load at a constant value. The pulsations have been removed. THE INDUCTOR OR CHOKE OPPOSES ANY CHANGE IN CURRENT. As a result, the output appears reasonably even and constant with very little RIPPLE or variation. Refer to waves in Fig. 8-32 and note filtering action.

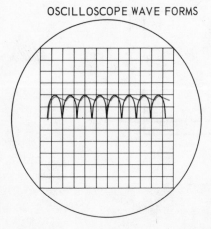

Fig. 8-32. Output wave form before (in black) and after filtering (in color).

AUDIO WAVE TO SOUND WAVE

An extremely important principle in electronics is the changing of an electrical wave, called the SIGNAL, to a sound wave which can be heard by the human ear. Your ears respond to sound waves. These waves come to your ear by vibrations of the air. The physicist describes a sound wave as alternate "condensations and rarefactions" of the air. To understand these words, refer to Fig. 8-33.

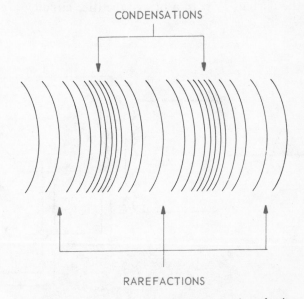

Fig. 8-33. Sound waves are condensations and rarefactions of air. Your ears respond to this sound wave.

An electronic device which changes electrical waves into sound waves is called a SPEAKER. Electrical waves can cause a diaphragm of paper or fiber to vibrate and produce sound waves. See Fig. 8-34. Here a small coil of wire, called a VOICE COIL, is suspended in a permanent magnetic field. Attached to the voice coil is a paper cone. A varying audio signal current connected to the voice coil will cause the coil to move outward and inward at the frequency of the audio wave. The cone makes the sound wave for you to hear. This is the principle of the PM (permanent magnet) SPEAKER.

Human hearing has its limitations. The range is generally considered between 20 and 20,000

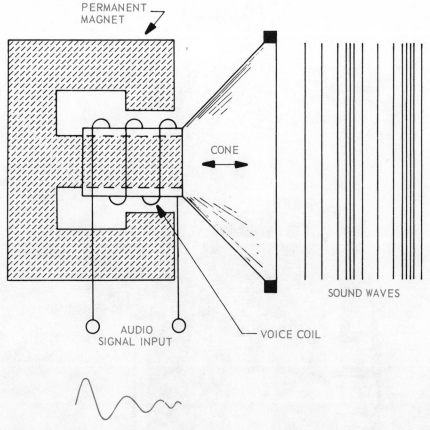

Fig. 8-34. In the PM speaker, the cone is caused to move in and out at the frequency of the input signal by the interaction between a permanent magnetic field and the electromagnetic field of the voice coil.

Hertz or cycles per second. Low tones have low frequencies; higher pitch tones have higher frequencies. Typical examples are:

Man's voice - average............125-130 Hz
Woman's voice - average.........250-260 Hz
Violin, up to approximately.........4,000 Hz
Middle A on piano...................440 Hz
Meadow Grasshopper singing.......14,200 Hz
Baby Robin tweeting..............15,000 Hz

The ability of a speaker to produce sounds of different frequencies depends on its size, construction and quality. A 3-way speaker in Fig. 8-35 contains a WOOFER for low frequencies, a TWEETER for the highs and an INTERMEDIATE SPEAKER for the in-between ranges. It is a HIGH FIDELITY SPEAKER.

MINI-SPEAKER PROJECT

The MS assembly plans are shown in Fig. 8-36. The completed speakers are shown in Fig.

8-37. The diagram and parts list for the MS is illustrated in Fig. 8-38. The speaker is easy to construct, and it has a wide range (70 to 16,000 Hz) output. Its design features a 4 in. acoustic-suspension woofer and a radial horn tweeter which gives great sound from an enclosure about one-third the size you would expect.

Refer to the diagram in Fig. 8-38 for details on wiring the two speakers. Also, note in Fig. 8-36 how the speaker is assembled.

FORWARD STEPS IN UNDERSTANDING ELECTRICITY-ELECTRONICS

1. A generator converts mechanical energy into electrical energy.
2. To induce a voltage, there must be a conductor, a magnetic field and relative motion between them.
3. Alternating current changes polarity and direction.

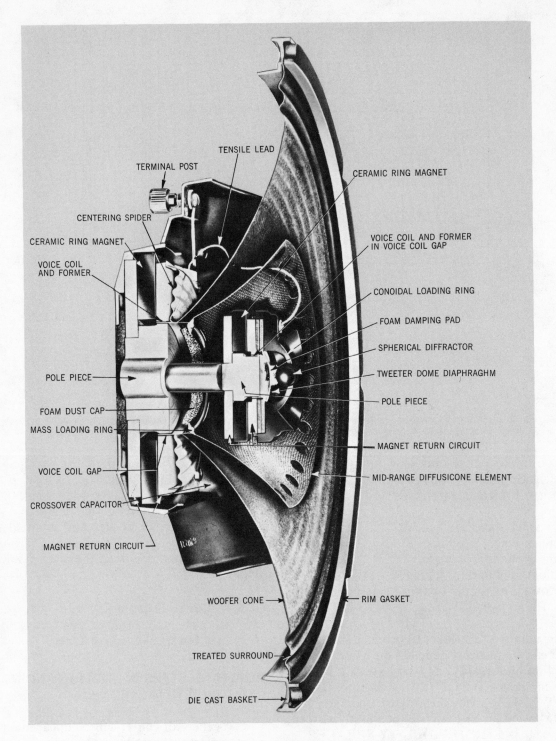

TENSILE LEAD

TERMINAL POST

CENTERING SPIDER

CERAMIC RING MAGNET

VOICE COIL
AND FORMER

POLE PIECE

FOAM DUST CAP

MASS LOADING RING

VOICE COIL GAP

CROSSOVER CAPACITOR

MAGNET RETURN CIRCUIT

CERAMIC RING MAGNET

VOICE COIL AND FORMER
IN VOICE COIL GAP

CONOIDAL LOADING RING

FOAM DAMPING PAD

SPHERICAL DIFFRACTOR

TWEETER DOME DIAPHRAGHM

POLE PIECE

MAGNET RETURN CIRCUIT

MID-RANGE DIFFUSICONE ELEMENT

WOOFER CONE

RIM GASKET

TREATED SURROUND

DIE CAST BASKET

Fig. 8-35. A high fidelity 3-way speaker.   (Universal Sound)

102

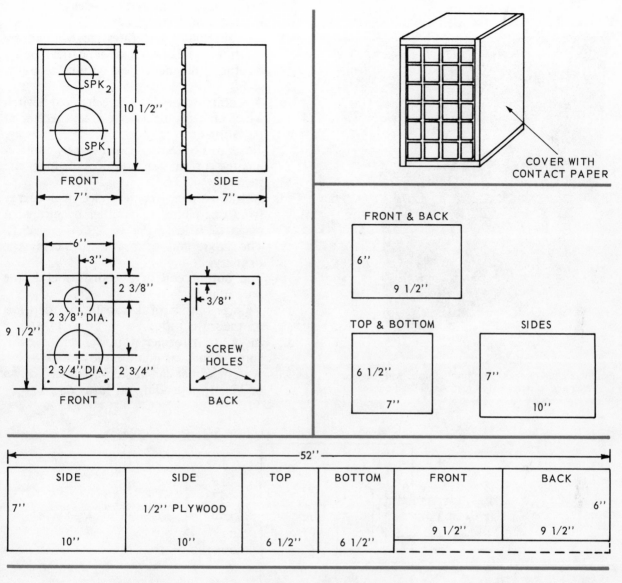

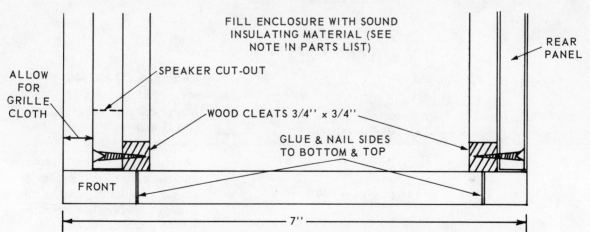

Fig. 8-36. Construction details for Mini-Speaker project.

Fig. 8-37. Mini-Speaker. In front: Speaker with sculptured foam grille removed. At rear: Assembled project.

4. Frequency is measured in cycles per second in units called HERTZ.
5. An alternator generates ac voltage by rotating a coil in a magnetic field or by rotating a magnetic field within the stator coils.
6. A commutator is a mechanical switch which changes the output of a generator to pulsating dc.
7. Generator output is controlled by controlling the current that produces the magnetic field.
8. A transformer transfers electrical energy from one circuit to another by means of magnetic fields.
9. The input coil of a transformer is the Primary.
10. The output coil of a transformer is the Secondary.
11. Voltage output of a transformer depends on the turns ratio.
12. In a transformer, "power in" equals "power out" assuming no losses.
13. A transformer can be step-up or step-down or 1 to 1, depending on turns ratio.

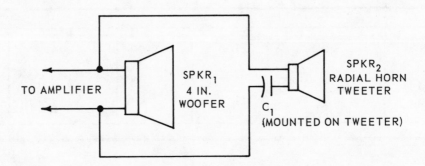

PARTS LIST FOR MINI-SPEAKER

SPKR$_1$ — 4 in. woofer, Radio Shack #40-1197 (70 to 15,000 Hz)

SPKR$_2$ — radial horn tweeter, Radio Shack #40-1278 (4000 to 16,000 Hz)

Enclosure — 1/2 in. plywood, 7 in. × 55 in. long

Cleats — 3/4 in. × 3/4 in. pine stock, 70 in. long

Grille — sculptured foam speaker grille, Radio Shack #40-1946

Misc. — wire, binding posts (2), wood grain contact paper, screws (wood and metal), paint, fiber glass or foam rubber

NOTE: Fill interior of speaker enclosure with 2-3 in. cubes of sound deadening material. Completely fill the enclosure, but do not pack tightly.

Fig. 8-38. Wiring diagram and parts list for Mini-Speaker.

14. Transformers are used by power companies to step up voltage for power transmission to decrease line losses.
15. A DIODE conducts electricity in only one direction.
16. Rectification means changing ac to dc.
17. A half-wave rectifier uses one-half the input cycle to produce a useful output.
18. A full-wave rectifier uses the complete input cycle to produce a useful output.
19. A rectifier produces a pulsating dc voltage. A filter is needed to remove pulsations.
20. Capacitors tend to keep the output voltage at a constant value.
21. Chokes or inductors tend to keep the output current at a constant value.
22. A PM speaker uses a permanent magnet.
23. A speaker converts ac waves at audio frequency into sound waves by interaction of magnetic fields.
24. The signal to a speaker is connected to the voice coil.

RULES OF SAFETY

THREE-WIRE CABLE

Receptacles, wires and appliances which meet modern standards of safety have a third or ground wire. A 3-prong plug is used. This is for your protection. Be certain that tools you use, such as drills, sanders and other machines, have their cases or frames connected to the ground.

SHARP METAL AND CORNERS

Small but painful cuts and bruises can happen when constructing metal chassis and cabinet enclosures. File all the sharpness from the edges. Remove any burrs made by drills and reamers. Always get FIRST AID for minor cuts. Report all injuries to your instructor.

TEST YOUR KNOWLEDGE - UNIT 8

1. A transformer has a 500-turn primary coil. Input voltage is 100V ac. What turns ratio will provide the correct number of secondary turns to give a final output of:
   1000V, 25V, 400V, 18V, 6V

2. If you connected dc voltage to a transformer, you would not get transformer action. Rather, the primary winding would heat up and destroy the transformer. Why?

3. How can an automotive alternator be used to charge an auto storage battery? Bear in mind that an alternator is an ac generator while the battery is dc.

4. Draw the circuit of a power supply which could be used for a 9 volt transistor radio.

5. A factory uses 10,000 watts of power. A 500,000 volt transmission line brings power from generator to distribution station, to factory. What current must flow in the 500 KV line to supply the factory?

6. Pay a visit to your local power company and find out for your area:
   1. Generator Source Voltage.
   2. High Tension Line Voltage.
   3. Distribution Line Voltage.
   4. City Branch Line Voltage.

7. If there is a power line near your home, answer these questions: Is there a transformer on it? What is the purpose of the transformer? How many homes on your street are served from one transformer?

8. Which is easier to filter, the output of a half-wave rectifier or a full-wave rectifier? Why?

9. A step-up transformer has a 100 volt input and a 500 volt output voltage. Which winding, the primary or secondary, will use the larger size wire for its coil?

Tire dynamics laboratory scientist "talks" to computer through keyboard to program test functions or interrupt tire tests already in progress. Programs stored on discs instruct computer for different tests.   (Firestone Tire & Rubber Co.)

Unit 9

ELECTRONIC COMPONENTS IN ACTION

There are many different types of electronic components used in circuits today. This Unit will discuss the following basic components:

1. What resistors are, and how they operate.
2. How capacitors work.
3. What factors make the inductor operate.
4. What impedance is.
5. What symbols and units are used for resistors, capacitors and inductors.

READING DIAGRAMS

In your studies of electronics, you will want to build many interesting and useful circuits. The major purpose of this book is to open the doors to exciting adventures in electronic science. You may wish to continue on to ad-vanced studies and a rich and rewarding career in electronics.

The skeleton, the backbone, the plan of all electronic circuits is the SCHEMATIC DIAGRAM. Consider that the carpenter, about to build a house, must first secure the architec-tural plans to find the sizes of rooms and their arrangement. The electronic technician, in turn, must secure a plan to find the size and arrange-ment of components such as resistors, capacitors and inductors. The components and devices must be connected together to produce the desired effects.

To learn to read and understand electronic components is a very important part of your studies. Fig. 9-1 is a schematic diagram of a sim-ple transistor radio. The components shown are made in a wide variety of sizes and values.

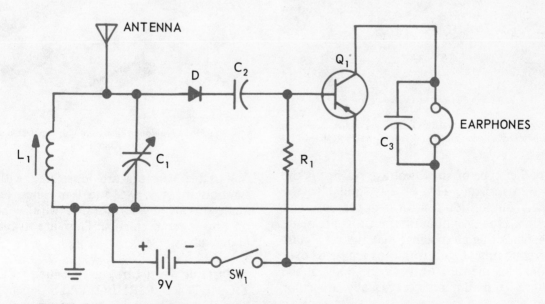

Fig. 9-1. A simple schematic diagram of a one-transistor radio. Note L, C and R components.

RESISTORS

You have already studied the effect of resistance in a circuit in Units 3 and 4. Resistors may be connected in series or in parallel. In either case, RESISTANCE is an opposing force to the flow of electricity. When a current flows through a resistor, the energy used creates heat. In your electric range at home, resistance is used to produce heat for cooking. On the other hand, heat produced by current through resistance in your Hi-Fi amplifier could be destructive unless provisions were made to radiate this heat into the surrounding air.

The common type of resistor is the carbon resistor shown in Fig. 9-2. The symbol is also shown. Carbon resistors are made in a wide variety of values of 2.7 ohms to millions of ohms. You will note a difference in physical size. The larger sizes have a greater radiation surface to dissipate the heat. These resistors can be purchased in 1/4 watt, 1/2 watt, 1 and 2 watt sizes.

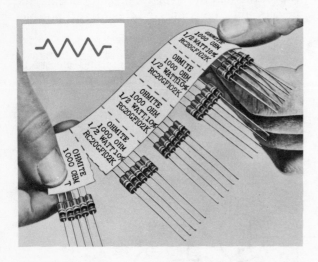

Fig. 9-2. Fixed carbon resistors. They are made in a wide variety of ohmic values and several wattage sizes.

Another type of small wattage resistor is the thin film resistor. This type is similar to the molded composition resistor in appearance and function. However, thin film resistors are made from depositing a resistance material on a glass or ceramic tube. Leads with caps are fitted over each end of the tube to make the body of the resistor. Thin film resistors usually are color coded.

For higher currents which produce more heat, a larger resistor is made by winding resistance wire on a ceramic form. These are called POWER RESISTORS, Fig. 9-3. They are made in 1, 2, 5, 10, 25, 50 and 100 watt sizes and even larger.

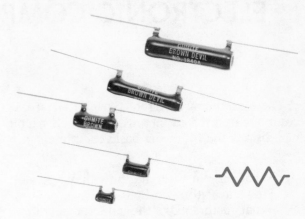

Fig. 9-3. Wirewound power resistors.

A variation of the power resistor has a slider on the body of the resistor which can be adjusted for some required resistance value. The adjustable power resistor and its symbol are shown in Fig. 9-4.

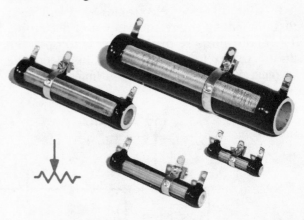

Fig. 9-4. Wirewound resistors with adjustable slider. (Ohmite)

Variable resistors are illustrated with their symbols in Figs. 9-5 and 9-6. These may be made of carbon or resistance wire, depending on power requirements. They are available in many values.

Variable resistors are called POTENTIOMETERS or RHEOSTATS depending on how they are used. Such a component is used

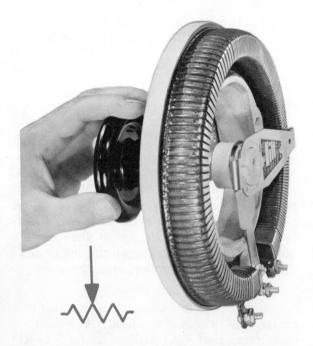

Fig. 9-6. Various types of potentiometers.   (Centralab)

Fig. 9-5. Wirewound rheostat for variable resistance in power circuits.   (Ohmite)

for the volume control on your radio. In Fig. 9-7, potentiometers are used in a television circuit to adjust height and width of picture.

CAPACITORS IN ACTION

The CAPACITOR is a new component in our studies. It will require some thought. A capacitor consists of two conducting plates separated by an insulator or nonconducting material. There is no electrical path through a capacitor for direct current. Follow the circuit in Fig. 9-8. Electrons flow to plate A. Because plate B is so close to plate A, the electrons are forced from plate B and are attracted to the positive terminal of the battery. The CAPACITOR becomes CHARGED to a polarity opposing the battery. Current flows in the circuit for only a fraction of a second while C is being charged. The capacitor now is a source of power in its own right. It can be removed from the circuit and used to light a

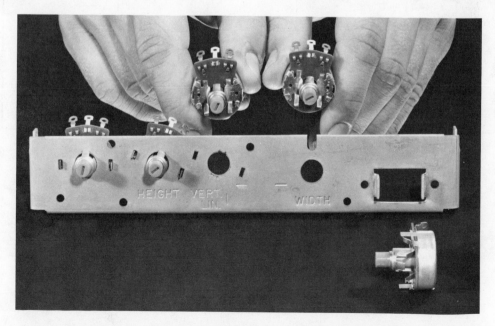

Fig. 9-7. Potentiometers are used on this television chassis for adjustments to picture height, linearity and width. These adjustments are made by the serviceman. They are usually found on the backside of a TV.   (Centralab)

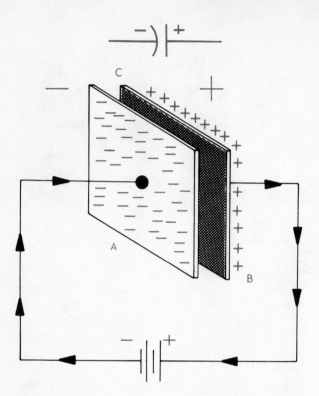

Fig. 9-8. Electrons flow from source to plate A. Electrons are repelled from plate B and are attracted to positive terminal of battery. Capacitor becomes charged.

small lamp. HIGHLY CHARGED CAPACITORS CAN DELIVER SERIOUS SHOCKS AND BURNS, IF IMPROPERLY HANDLED.

CAPACITORS AND AC

When connected to an alternating current source, a capacitor performs in a different manner. The capacitive circuit is redrawn, using the symbol shown in Fig. 9-9. When the ac is on its positive half-cycle, the capacitor charges to one polarity. When the ac is on its negative half-cycle the capacitor charges to the opposite polarity. Current is flowing in the circuit at all times; first in one direction, then the other.

You may raise this question: How much of a charge can a capacitor hold? The unit of measurement of CAPACITY is the FARAD (F), which means it will hold a specified number of electrons when one volt is applied. In electronics, smaller units for measuring capacity are used. They will be marked in microfarads (μF) and picofarads (pF).

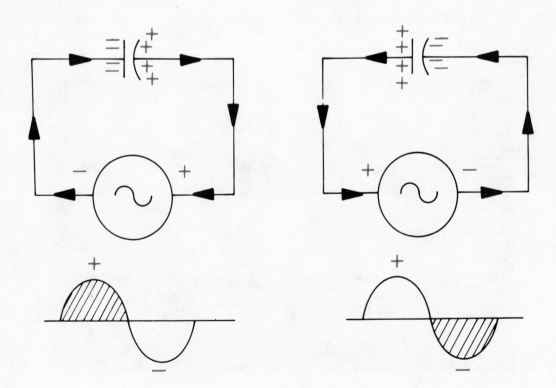

Fig. 9-9. The capacitor charges and discharges according to the ac voltage source.

A microfarad is 1/1,000,000 of a farad. A picofarad is 1/1,000,000 of 1/1,000,000 of a farad. A chart that shows the relationship of the microfarad and the picofarad to the basic farad unit is shown in Fig. 9-10.

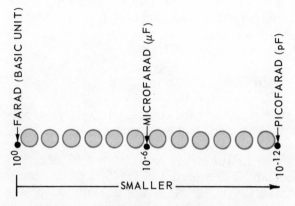

Fig. 9-10. Prefix chart for capacitance units.

Also worth noting, a capacitor will have a WORKING VOLTAGE RATING, and it will be so marked. If this voltage is exceeded, the capacitor will short out and be destroyed.

EXPERIENCE 1. Connect a 50 μF, 50 volt capacitor in series with a 6 volt lamp to your 6 volt dc supply. See Fig. 9-11. Does the lamp glow?

CONCLUSION: A capacitor blocks dc and the lamp will not glow.

EXPERIENCE 2. Use the same circuit but connect it to a 6 volt ac power source. Does the lamp glow? Does it glow as brightly as it should?

CONCLUSION: Yes, the light does glow when connected to ac. There is ac in the circuit due to the charge and discharge of C. But it does not glow brightly. The capacitor is not large enough. It has insufficient capacity. Not enough current is allowed to flow.

EXPERIENCE 3. Connect another 50 μF capacitor in parallel to C_1. Refer to the circuit in Fig. 9-12. Now the lamp glows much brighter.

CONCLUSION: The capacitance of the circuit is now twice as much and the charging current causes the lamp to glow brighter.

Thinking about these last two experiences, you might come to another conclusion. The single capacitor permitted some current to flow. Increasing the capacitance by adding C_2 permitted more current to flow. That sounds like resistance by another name. This is true. The

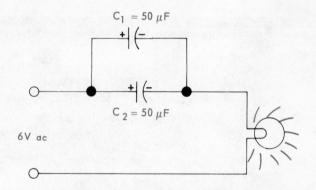

Fig. 9-12. A second capacitor C_2 is connected in parallel to C_1. The capacitance has doubled and the current has increased.

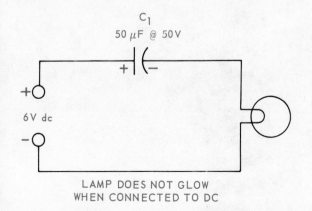

LAMP DOES NOT GLOW
WHEN CONNECTED TO DC

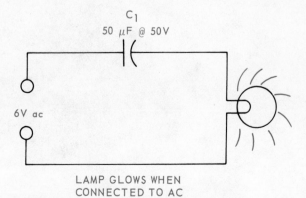

LAMP GLOWS WHEN
CONNECTED TO AC

Fig. 9-11. A capacitor blocks dc but permits some ac to flow to light the lamp.

opposition of a capacitor to the flow of alternating current is called REACTANCE. It, too, is measured in ohms, like resistance. Reactance in ohms depends on the size of the capacitor and the frequency of the alternating current.

TIME CONSTANTS

An extremely useful capacitive action is the time required to charge a capacitor. Look over the circuit in Fig. 9-13. The capacitor is connected in series with a resistor to the 15 volt dc

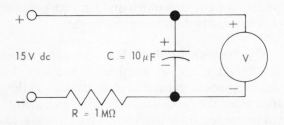

Fig. 9-13. When power is turned on, C charges slowly as indicated on the voltmeter.

power source. When the circuit is turned on, the capacitor cannot charge in an instant. The resistor limits the electron flow. Some time must pass before C becomes charged to about full voltage.

The TIME CONSTANT of the circuit is the time required for C to charge to about 63 percent of its final value. This interval of time can be computed by the formula:

$$R \text{ in ohms} \times C \text{ in farads} = T \text{ in seconds}$$

At the end of 5 time constant periods the capacitor is assumed to be charged.

EXPERIENCE 1. A circuit of a 10 μF capacitor and a 1 megohm (1,000,000 Ω) resistor is connected in series to the 15 volt dc power source, in Fig. 9-13. To watch C charge, a voltmeter (VTVM) is connected across its terminals. Turn on the power and observe the movement of the meter.

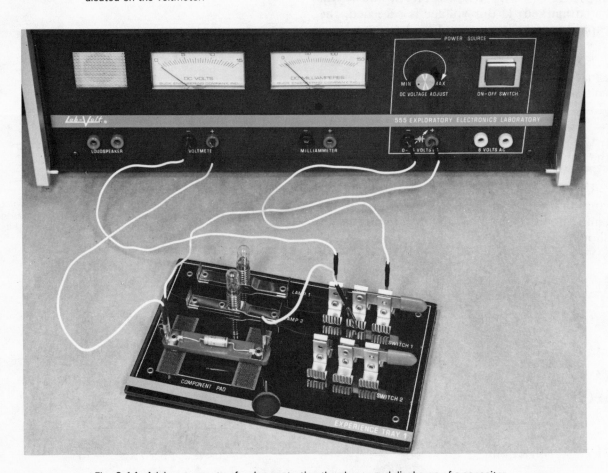

Fig. 9-14. A laboratory setup for demonstrating the charge and discharge of a capacitor.

EXPERIENCE 2. After C is charged in EXPERIENCE 1, disconnect the two leads to the power source and connect the two leads together. C will now slowly discharge as indicated by the meter.

CONCLUSIONS: The rate at which a capacitor charges and discharges depends upon the resistance and capacitance of the circuit. A laboratory setup for demonstrating the charge and discharge of a capacitor is shown in Fig. 9-14.

THE CAPACITOR FAMILY

A variable capacitor is illustrated in Fig. 9-15. Note the plates and how they mesh together when the rotor is turned. In the fully meshed position, there is maximum capacitance. The insulation between the plates is AIR. In Fig. 9-16, a small variable ceramic trimmer capacitor is illustrated. It is adjusted by a screwdriver.

Fig. 9-17 pictures the common paper capacitor. The plates are tinfoil separated by waxed paper. It is rolled up in tubular form and coated

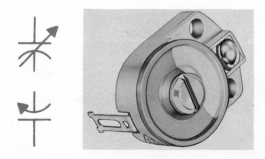

Fig. 9-16. A small adjustable TRIMMER capacitor. Used for only small changes in capacitance. These adjustments are made by a technician when your equipment needs service. (Centralab)

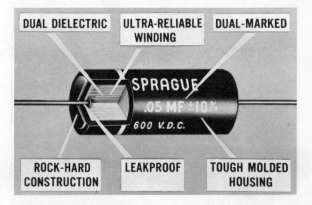

DUAL DIELECTRIC   ULTRA-RELIABLE WINDING   DUAL-MARKED

SPRAGUE .05 MF ±10% 600 V.D.C.

ROCK-HARD CONSTRUCTION   LEAKPROOF   TOUGH MOLDED HOUSING

Fig. 9-17. A molded paper capacitor.

with wax or plastic. These are made in many values and voltage ratings. Get a paper capacitor from an old radio or TV. Cut it open with a knife and see how it is made.

Dozens of small ceramic disc capacitors, Fig. 9-18, can be found in a TV set. These are used where small values of capacitance are required. A ceramic disc, which is the insulator, has silver deposited on each side for plates. Wire leads are attached to the plates and the whole device is coated with a protective material. Capacitors are measured in picofarads and microfarads.

When a large amount of capacitance is needed and there is limited space, an ELECTROLYTIC capacitor is useful. A tubular type appears in Fig. 9-19. These are formed by a strip of aluminum coated with oxide and a layer of paper or gauze soaked in a paste-like electrolyte. A metal contact plate completes the sandwich. It is then rolled up in tubular form

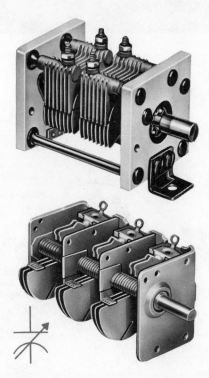

Fig. 9-15. Variable capacitors are manufactured in many sizes and types to fill the requirements of electronic circuits. (J. W. Miller Co., and Hammarlund Mfg. Co.)

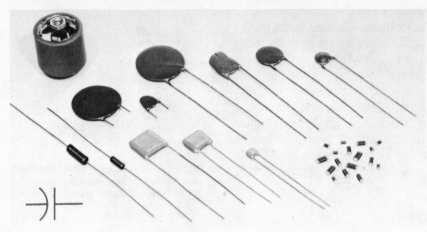

Fig. 9-18. Ceramic capacitors.   (Centralab)

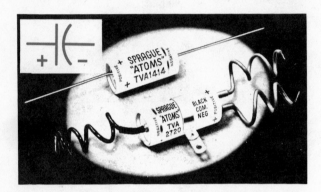

Fig. 9-19. A tubular type electrolytic capacitor.

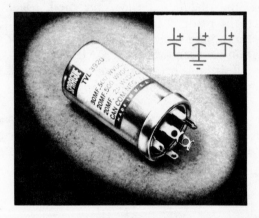

Fig. 9-20. A can type electrolytic capacitor. It can be rigidly mounted on a chassis.

like the paper capacitor. These electrolytic capacitors usually have values from 1 to 1000 or more microfarads and many working voltage ratings.

It is important to remember that the ELEC-TROLYTIC CAPACITOR HAS A PLUS AND MINUS POLARITY. IT MUST BE IN-STALLED CORRECTLY. Failure to do so will ruin the capacitor, (sometimes they spark and smoke) and your circuit may not work.

A can type electrolytic capacitor is shown in Fig. 9-20. These are mounted on the top of a chassis. Some cans contain three or four capacitors of different values.

A popular electrolytic capacitor used in transistor circuits and on etched circuit boards is the tiny, low voltage electrolytic capacitor pictured in Fig. 9-21.

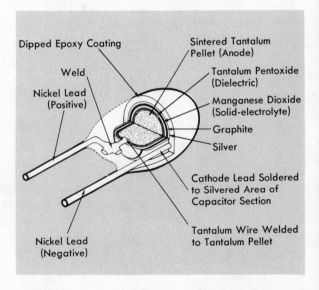

Fig. 9-21. Hermetically sealed, solid-electrolyte tantalum electrolytic capacitor.   (Sprague)

COILS AND INDUCTORS

In your studies of magnetism, you have already come face-to-face with the solenoid coil and electromagnet. Coils are also used for many other purposes. Now we will investigate the action of a coil in a circuit.

EXPERIENCE 1. A coil is connected in series with a 6 volt lamp to a 6 volt dc power supply in Fig. 9-22. The lamp glows at full brightness.

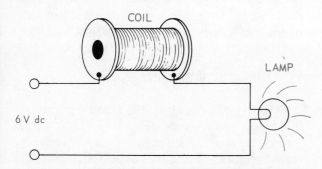

Fig. 9-22. The coil and lamp are connected to a dc source. The lamp burns brightly.

CONCLUSION: The coil seems to have little effect upon the flow of direct current. The only resistance in the circuit is the resistance of the wire in the coil.

EXPERIENCE 2. Connect the same coil and lamp to a 6 volt ac power supply. The lamp now glows dimly. Refer to Fig. 9-23.

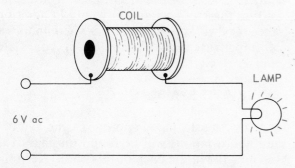

Fig. 9-23. The coil and lamp are connected to an ac source. The lamp burns dimly.

CONCLUSION: The coil must oppose the flow of an alternating current. That is true. This opposition to ac is called INDUCTIVE REACTANCE and is measured in OHMS just like resistance.

EXPERIENCE 3. Continue EXPERIENCE 2 by placing a round iron core in the coil. See Fig. 9-24. Now the lamp burns very dimly or not at all. Move the core in and out and note how the brightness of the light will change.

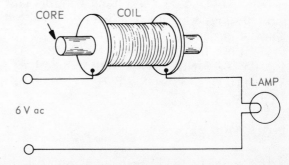

Fig. 9-24. With an iron core placed in the coil, the lamp is very dim or not burning at all.

CONCLUSION: By placing the core in the coil, the coil MUST HAVE MORE REACTANCE. Much less current now flows.

COUNTER ELECTROMOTIVE FORCE (CEMF)

You will recall that another name for voltage is electromotive force or EMF. Voltage causes the current to flow in a complete circuit. Counter electromotive force or CEMF is a counter voltage developed which opposes the source voltage. How can that happen?

First, study Fig. 9-25. Note that a coil is connected to an alternating current supply. Consider the first half-cycle. As the applied voltage increases in the positive direction, the MAGNETIC FIELD also increases in strength

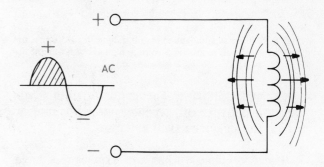

Fig. 9-25. The rising voltage and current causes an expanding magnetic field that cuts across the coil windings and induces a counter voltage of opposite polarity.

and expands outward. The magnetic field expanding outward cuts across the windings of the coil itself and INDUCES A VOLTAGE WHICH OPPOSES THE SOURCE VOLTAGE.

Now, you should understand why the lamp glowed dimly when the coil is in series with it. Current in a circuit depends upon voltage. THE CURRENT IN THIS CIRCUIT IS PRODUCED BY THE NET VOLTAGE OR SOURCE VOLTAGE MINUS THE COUNTER VOLTAGE.

INDUCTIVE KICK

If a magnetic field is expanding and collapsing at a rate of 60 times per second, a certain CEMF will be developed. If the rate is speeded up to 1000 Hz, then a greater CEMF will be developed. For example, increasing the revolutions per minute of a generator will increase the rate at which the conductors cut across the magnetic field. Therefore, a greater output.

EXPERIENCE 4. The circuit in Fig. 9-26 should be made. Note that the coil is energized when you push the button switch. When the switch is released, the magnetic field collapses and produces a voltage high enough to light the NE2 neon lamp. This lamp requires 65 volts to light it.

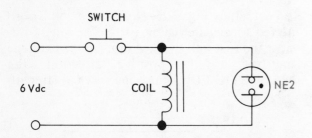

Fig. 9-26. The rapid collapse of the magnetic field when the switch is opened produces a voltage high enough to light the NE2 glow lamp.

CONCLUSIONS: The amount of CEMF developed by a coil depends upon the rate at which the magnetic field collapses.

The foregoing studies and experiences have many applications in electronics and industry. When an inductive circuit is opened either by a switch or a relay, high voltages are induced across the switch contacts. The "arcing" that results will burn up the switch contacts and cause interference to radio communications nearby.

HENRY (H)

When a coil or inductor appears in a circuit, it is marked with the letter L. Inductance is measured in HENRYS. For your information, if a coil produces one volt of CEMF when the current through the coil changes at a rate of one ampere per second, the coil is said to have an inductance of ONE HENRY. Coils used in electronic work will have values in henrys, millihenrys (mH) and microhenrys (μH).

THE INDUCTOR FAMILY

Typical coils found in electronic circuits are illustrated and named in Fig. 9-27. Their names suggest their use. Notice that many of them are transformers. This is expected, since a transformer is just another way of using coils. There are thousands of types and values in common usage. Some you will use in interesting circuits and experiments later in this text.

IMPEDANCE (Z)

Many circuits contain resistance, capacitance and inductance. Each has its own way of opposing the flow of alternating current. The total effect is called the IMPEDANCE (Z) of the circuit. It is also measured in ohms. You cannot add up these effects as you would in addition, since they work in different ways. Computation of impedance will be studied in your later courses in electronics.

SIGNS AND SYMBOLS

A complete listing of approved signs, symbols and abbreviations will be found on pages 188, 189 and 190 in the back of this book. Keep this in mind for ready reference.

IC DECISION MAKER PROJECT

Have you ever wanted someone to make a decision for you? This electronic device, Fig. 9-28, can do just that! By pressing the push button switch, you have a fifty-fifty chance on

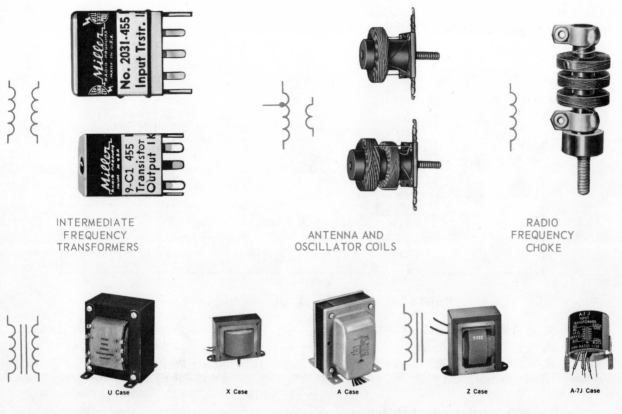

Fig. 9-27. Typical inductors and transformers used in electronic circuits.

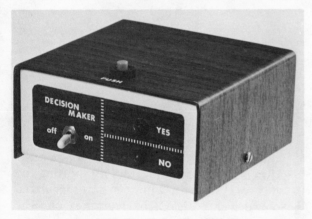

Fig. 9-28. The IC Decision Maker.

whether the "YES" or "NO" light emitting diode stays on.

Refer to Fig. 9-29 for the schematic and parts list for the decision maker. The wiring should be straightforward with no tricks. It is a good idea to use sockets for the ICs and to construct a printed circuit for the project.

FORWARD STEPS IN UNDERSTANDING ELECTRICITY-ELECTRONICS

1. A capacitor consists of two conducting plates separated by an insulator.
2. Capacitance is measured in farads, microfarads and picofarads.
3. The opposition to an alternating current by a capacitor is called CAPACITIVE REACTANCE. It is measured in OHMS.
4. A capacitor blocks direct current.
5. Resistors are rated in watts, which means the ability to dissipate heat.
6. A "potentiometer" is a variable resistor.
7. An inductor is a coil of wire.
8. Inductance is measured in henrys, millihenrys and microhenrys.
9. Opposition to ac by an inductor is called INDUCTIVE REACTANCE.
10. Capacitance is the property of a circuit to oppose a CHANGE IN VOLTAGE.
11. Inductance is the property of a circuit to oppose a CHANGE IN CURRENT.

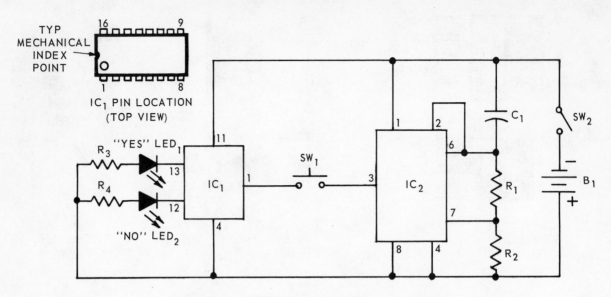

PARTS LIST FOR THE IC DECISION MAKER

R₁ —10 KΩ, 1/2 W resistor
R₂ —47 KΩ, 1/2W resistor
R₃, R₄ —330 Ω, 1/2W resistors
C₁ —capacitor, .01 μF @ 50 WVdc
IC₁ —integrated circuit, TTL flip flop, SN 7473N or HEP C7473P
IC₂ —integrated circuit, LM 555 or NE 555V

B₁ —5V power supply or 3 - AA penlight cells
SW₁ —push button switch, NO
SW₂ —SPST toggle switch
LED₁, LED₂ —light emitting diodes, XC 556Y or equivalent
Misc. —PC materials and board, case, wire, solder, LED mounts, IC sockets (8 pin, 16 pin)

Fig. 9-29. Schematic and parts list for the IC Decision Maker.

12. The time constant of an RC circuit is the resistance in ohms times capacitance in farads equals time in seconds.

RULES OF SAFETY

CHARGED CAPACITORS DESTROY METERS

High voltages stored in capacitors may destroy a meter, especially ohmmeters, if measurements are being made. Always discharge capacitors to ground by touching an INSULATED SCREWDRIVER between the capacitor and the chassis ground, BEFORE servicing a circuit. Also discharge capacitors by shorting their terminals with an insulated wire.

CAPACITORS STAY CHARGED

Many capacitors remain charged after the equipment is "turned off." Some retain a voltage which could knock you down. Always discharge capacitors by shorting the ends together with a jumper or an insulated screwdriver, BEFORE working on the circuit.

RESISTORS ARE HOT COMPONENTS

Current flowing through resistance does work and produces HEAT. If your electronic equipment has been on for a while, some of the resistors get very hot. Watch out when working on the circuit. You can really burn your fingers.

Unit 10

VACUUM TUBES AND TRANSISTORS

The first two generations in the family of electronics were the vacuum tube and the transistor. In this Unit, you will study:

1. How the vacuum tube operates.
2. What the transistor is, and how it works.
3. What bias is.
4. How vacuum tubes and transistors can rectify and amplify.
5. What the light emitting diode is, and how it operates.
6. How some basic transistor amplifier circuits work.

WIZARD AT WORK

We are truly indebted to the genius of a great American scientist, Thomas Edison. Early in his experimentation, he discovered that an electric current flowing through a resistance wire would produce heat. How hot could he get the wire? Red hot? White hot? Hot enough to produce a glowing light?

The problem was to discover some wire which would not burn up if hot enough to produce light. A really hot wire would oxidize with the surrounding air and destroy itself. The first successful incandescent lamp was made with a thread from Mrs. Edison's sewing basket coated with carbon and placed in a vacuum bottle. It burned for 40 hours. Lamps of today use tungsten wire for the filament and industry has found better ways of producing a vacuum.

The story does not end there. Edison continued his experimentation with "the glowing lamp in a bottle." Little wonder he was called

the "Wizard of Menlo Park." Edison wondered if an electric current would flow through a vacuum. His original experiment was set up similar to Fig. 10-1. A voltage called the A supply was connected to heat the light filament. A second voltage known as the B supply was connected to make an independent circuit between the filament and a PLATE of conducting metal in the lamp.

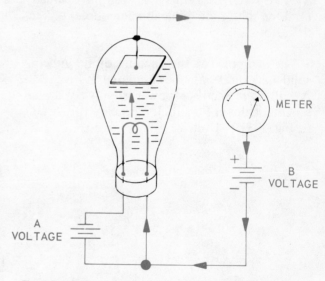

Fig. 10-1. Edison's experiment which demonstrates an electric current can flow through a vacuum.

THE DIODE — A ONE-WAY STREET

In Edison's first experiment, he placed the B battery so its NEGATIVE terminal connected to the plate. Nothing happened. When he turned around the B battery so its positive terminal connected to the plate, a current through the tube was indicated on the meter.

Edison had produced a DIODE, (A ONE DIRECTION ONLY VALVE). Current would flow from filament to plate but not from plate to filament. The British still call this device a VALVE. The theory of operation of this diode remained a mystery for some years. At that time, electric current was not known as "electron flow."

In truth, the ELECTRON THEORY was not proposed until 1900 by J. J. Thomson. The explanation of this phenomenon seems quite commonplace today. When Edison heated the filament, electrons "boiled" out and formed a cloud of electrons around the filament. This is called the Edison Effect. When the plate was "positive," it attracted the electrons through the vacuum and caused current to flow. When the plate was negative, it repelled the electrons.

Through the years, diodes have had many applications as rectifiers and radio detectors. During that time, the vacuum diode was made obsolete by the development of the SOLID STATE DIODE. Remember, you have already used the solid state diode in your studies on rectification.

The symbols for the vacuum diode and the solid state diode may be compared in Fig. 10-2. A full-wave rectifier power supply is drawn schematically in Fig. 10-3. Compare this circuit to Fig. 8-31 in Unit 8.

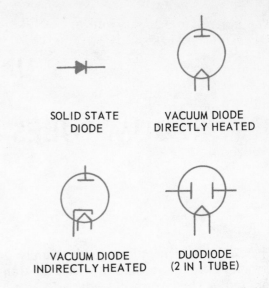

SOLID STATE DIODE   VACUUM DIODE DIRECTLY HEATED

VACUUM DIODE INDIRECTLY HEATED   DUODIODE (2 IN 1 TUBE)

Fig. 10-2. Symbols for solid state and vacuum diodes.

BIRTH OF ELECTRONIC AGE

At the turn of the last century, one of America's great scientists, Dr. Lee DeForest, continued experimentation with the diode valve tube. He inserted a screen of wire between the heated cathode (the cathode emits the electrons) and the plate of the tube. Its effect had such a tremendous impact on electronic science that DeForest is often referred to as the "father of modern radio." He named this screen the GRID.

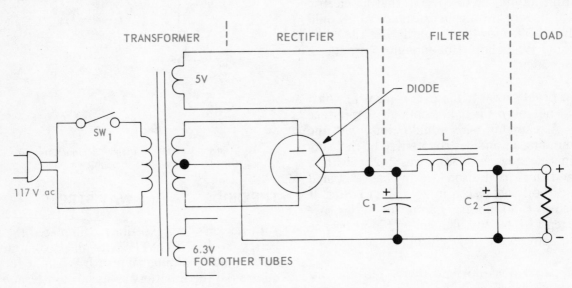

Fig. 10-3. A circuit for a full-wave rectifier power supply using a duodiode vacuum rectifier.

The symbol for a tube with a CONTROL GRID is shown in Fig. 10-4. It is called a TRIODE. Note that the grid is close to the cathode and all electrons must pass through the grid on their way to the plate. If the grid is made slightly NEGATIVE, it repels some of the electrons trying to pass through. Therefore, the electrons collected by the plate become less.

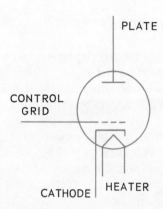

Fig. 10-4. A three element tube called the TRIODE. The heater is not counted unless it is also the cathode.

If the grid is made more NEGATIVE, only a few electrons find their way to the plate. The GRID can be made NEGATIVE enough to prevent ANY ELECTRONS from passing through to the PLATE. The tube is said to be CUT OFF. A graph in Fig. 10-5 illustrates this action. Note that as the grid becomes more negative, the plate current decreases until it becomes ZERO.

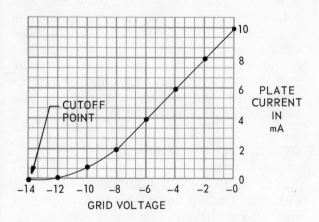

Fig. 10-5. Graph shows decrease in plate current as grid is made more negative.

AMPLIFICATION

The astounding fact in the operation of the TRIODE (three element) TUBE is this: A VERY SMALL CHANGE IN GRID VOLTAGE WILL PRODUCE A REMARKABLE CHANGE IN PLATE CURRENT.

A load resistor is connected in the plate circuit in Fig. 10-6. The plate current must flow through this resistor. Therefore, a voltage will appear across this resistor (E = I × R). Assume the load resistor is 25,000 ohms. One milliampere (.001 amp) of current would produce a voltage across the load resistor of:

$$.001 \text{ amp} \times 25,000 \ \Omega = 25 \text{ volts}$$

Referring back to the graph in Fig. 10-5, you will see that it requires ONLY ONE VOLT CHANGE IN GRID VOLTAGE TO PRODUCE A ONE MILLIAMPERE CHANGE IN PLATE CURRENT, WHICH CAUSES A 25 VOLT CHANGE ACROSS THE RESISTOR. THIS IS AMPLIFICATION. In this example, the voltage amplification is 25 times.

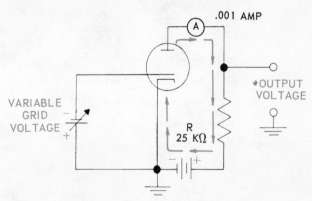

Fig. 10-6. A small change in grid voltage produces a large change in output voltage. This is voltage amplification.

BIAS VOLTAGE

In order to faithfully amplify a signal (ac voltage), we will select some fixed value of grid voltage and call this our OPERATING POINT or Q point. By doing this, the grid voltage can be varied up and down while the current can be made to vary around a fixed point. This fixed voltage is called the BIAS VOLTAGE.

In the circuit drawn in Fig. 10-7, we have selected −4 volts as the fixed BIAS. The current at this point is 6 mA. We also have an ac generator with a one volt peak output connected in series with the fixed bias voltage. Therefore, when ac generator output is zero, grid voltage is −4 volts. When ac generator output is +1 volt (peak), grid voltage is −3 volts:

$$(-4) + (+1) = -3$$

When ac generator output is −1 volt, grid voltage is −5 volts:

$$(-4) + (-1) = -5$$

As the ac generator causes the grid voltage to vary back and forth from −4 to −3 to −4 to −5, the current through the tube varies from:

$$6 \text{ mA to } 7 \text{ mA to } 6 \text{ mA to } 5 \text{ mA}$$

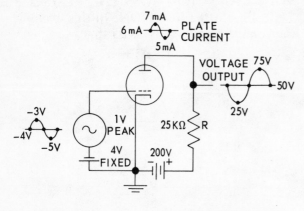

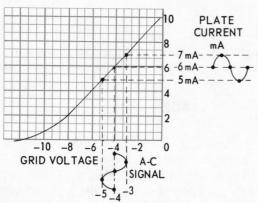

Fig. 10-7. A 2 volt peak-to-peak change at the grid produces a 50 volt peak-to-peak change at the plate.

The voltage across the load resistor varies from:

$$150V \text{ to } 175V \text{ to } 150V \text{ to } 125V$$

Therefore, we have an output of an ac wave which has a peak-to-peak value of 50 volts (25 to 75) as the result of an input voltage signal with a two volt peak-to-peak value (−3 to −5). The gain is 25 times.

OUTPUT VOLTAGE

The output voltage appears to be upside down compared to the input. However, this is always true. A vacuum tube INVERTS the input signal. You can reason this out:

In Fig. 10-7, at the fixed operating point, there are 150 volts across R. This leaves 50 volts of the total voltage to appear across the tube. At 7 mA of current, 175 volts appear across R but only 25 volts across the tube. At 5 mA of current, only 125 volts appear across R, leaving 75 volts across the tube. As a result, the voltage at the plate measured from ground is opposite in polarity to the input signal.

VACUUM TUBE AS A SWITCH

There are numerous opportunities for the use of a fast-acting switch in electronic circuitry. A vacuum tube can be used for this purpose.

According to the graph in Fig. 10-5, at negative 12 volts on the grid, the tube is cut off. No current flows. It is like an open switch with infinite resistance. On the other hand, with zero volts at the grid, the tube is approximately in full conduction. Therefore, a small positive voltage on the grid could trigger the tube into conduction. The plate current could activate a relay and thereby switch very high power circuits, motors and machinery. A schematic of a switching circuit is shown in Fig. 10-8.

OTHER VACUUM TUBES

Several types of vacuum tubes are illustrated with their symbols and BASE DIAGRAMS in Figs. 10-9 and 10-10. Some have special grids such as the SCREEN GRID and SUPPRESSOR GRID to improve their performance.

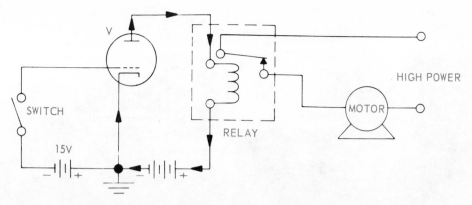

Fig. 10-8. A vacuum tube can be used as a switch for high power circuits.

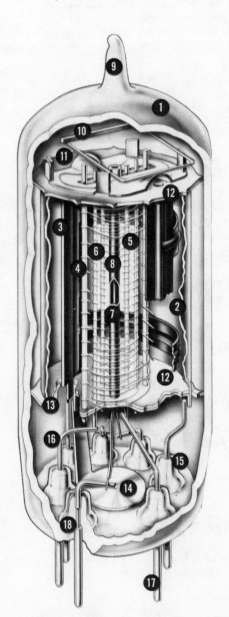

1—Glass Envelope

2—Internal Shield

3—Plate

4—Grid No. 3 (Suppressor)

5—Grid No. 2 (Screen)

6—Grid No. 1 (Control Grid)

7—Cathode

8—Heater

9—Exhaust Tip

10—Getter

11—Spacer Shield Header

12—Insulating Spacer

13—Spacer Shield

14—Inter-Pin Shield

15—Glass Button-Stem Seal

16—Lead Wire

17—Base Pin

18—Glass-to-Metal Seal

Fig. 10-9. The internal structure of a miniature tube.

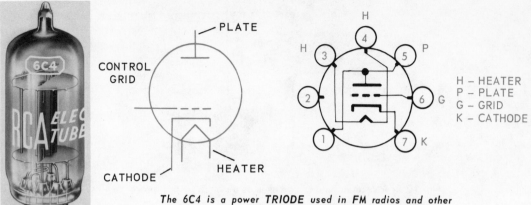

The 6C4 is a power *TRIODE* used in *FM* radios and other high frequency circuits.

H – HEATER
P – PLATE
G – GRID
K – CATHODE

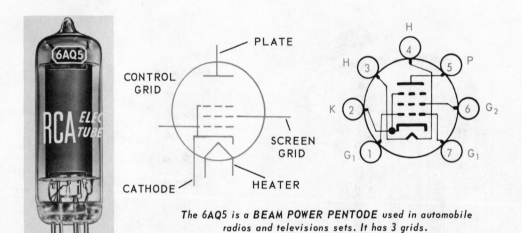

The 6AQ5 is a *BEAM POWER PENTODE* used in automobile radios and televisions sets. It has 3 grids.

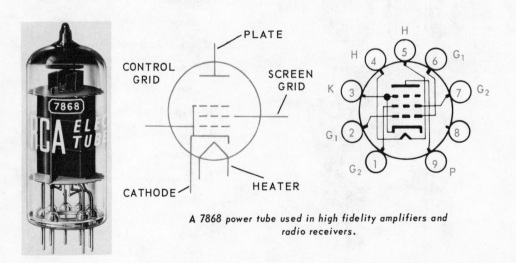

A 7868 power tube used in high fidelity amplifiers and radio receivers.

Fig. 10-10. Several types of vacuum tubes are pictured with their symbols and base diagrams.

Tubes serve a wide variety of applications: miniature tubes for radio frequency signals; large tubes handle high power; some tubes amplify voltage, others amplify current. Read the descriptions of these typical tubes in the illustrations.

THE TRANSISTOR

The invention of the transistor has revolutionized the electronics industry. They are tiny devices, yet very rugged and dependable. They require "no heat for warm-up" like the vacuum tube. By using transistors, we have many types of light, portable electronic equipment. Space travel to the Moon and sophisticated computers were made possible by the transistor. The future of electronics is in the world of SOLID STATE DEVICES. The three scientists pictured in Fig. 10-11 received the Nobel Prize in Physics for their discovery of the transistor effect in 1948.

Fig. 10-11. The transistor inventors (left to right) are Dr. William Shockley, Dr. Walter Brattain and Dr. John Bardeen. All were engineer-scientists of the Bell Laboratories in 1948. (Bell Telephone Laboratories)

SOLID STATE CRYSTALS

Transistors and other solid state devices are mostly made of crystals of germanium and silicon, although other elements are used. These crystals in their pure form are very poor conductors of electricity. The crystals are changed by adding a very small quantity of an impurity that makes the crystal a SEMICONDUCTOR. It is neither a good nor a bad conductor.

ELECTRON CARRIERS

In one case, an impurity is added which causes some free electrons to become available

for conduction. This crystal is an N type crystal. Current flows through it in much the same manner as in conductors we have already discussed. An N crystal is displayed in Fig. 10-12. The current carriers are ELECTRONS.

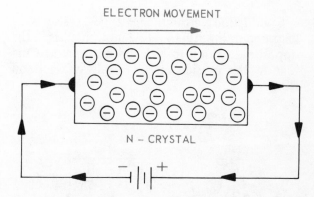

Fig. 10-12. An N type crystal conducts by electron movement like the typical conductor.

HOLE CARRIERS

In the second case, the impurity added to the crystal produces HOLES. A HOLE can be defined as a POSITIVE LOCATION which has a strong attraction for an electron. These are called P type crystals. Current also can be conducted by HOLES. Study Fig. 10-13. An electron drawn to the positive power source leaves a

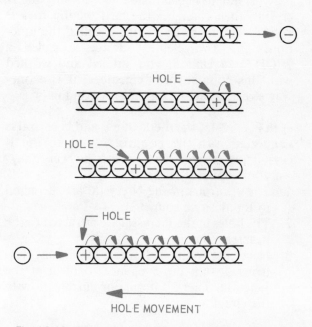

Fig. 10-13. Conduction in P crystal by HOLE movement.

vacancy. The neighboring electron jumps over to the hole and leaves a vacancy. The next electron jumps into that vacancy and leaves a hole in turn. The HOLE, as you will observe, moves from right to left, where it is finally filled by an electron from the negative power source.

Conduction through a P type crystal is illustrated in Fig. 10-14. Hole movement is in the opposite direction to electron movement. Current in the outside circuit is always electron flow.

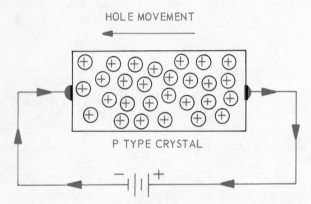

Fig. 10-14. In the P type crystal, the current carriers are HOLES. Electrons flow in external circuit.

ELECTRONS MEET HOLES

A crystal device is made by forming an N crystal, then changing the next section to a P crystal. The point of change is called the junction. This two element device is called a DIODE. In Unit 8, you studied and worked with this component. Remember, it is a one-way street; a one-direction only conductor.

In Fig. 10-15, the diode of a P and N crystal is connected with the negative source to the P crystal and the positive source to the N crystal. Very little current flows:
1. The electrons in the N crystal are attracted to the positive source.
2. The holes in the P crystal are attracted to the negative source.
3. Never the twain shall meet. Holes and electrons do not get a chance to meet at the junction. There is no major current flow in the circuit.

This is stated as being REVERSE BIASED.

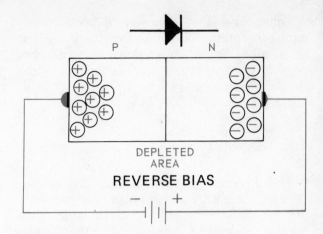

Fig. 10-15. Crystal diode connected in a REVERSE BIAS direction. No major current flow.

Now, we will connect the diode in a FORWARD BIAS direction as in Fig. 10-16. The battery has been turned around: negative terminal to the N crystal and positive terminal to the P crystal.

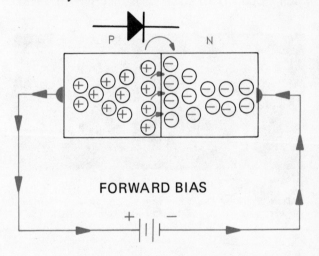

Fig. 10-16. Crystal diode connected in a FORWARD BIAS direction. Major current flows.

In this connection:
1. Electrons move from the source, through the N crystal to the junction.
2. At the junction, the holes in the P crystal are filled with electrons.
3. The electrons, attracted by the positive power source, create more holes in the P crystals.
4. The new holes move toward the junction to be refilled.
5. Current flows in the circuit; electrons flow in the external circuit.

126

EXPERIENCE 1. Select a solid state diode from your electronics parts. Using the ohm-meter, measure the forward and reverse resistance of the device.

CONCLUSION: The diode must limit current to a very small amount in the reverse direction because its resistance is very high. The low resistance in the forward direction permits current to flow.

EXPERIENCE 2. Connect the diode, resistor and milliammeter according to the schematic in Fig. 10-17. Measure the current with the diode in forward bias connection. Remove the diode from the circuit and reconnect it in a reverse bias connection. Observe the current reading.

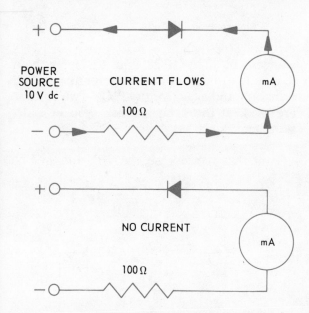

Fig. 10-17. Current flows only when diode is connected in forward bias. Resistor limits the current.

CONCLUSION: Current flows easily in the forward bias direction, but very little current flows in the reverse direction.

Diodes are manufactured in a wide range of sizes and current and power ratings. A typical TOP-HAT silicon diode is illustrated in Fig. 10-18.

LIGHT EMITTING DIODES (LED)

The light emitting diode (LED) is a "solid state" indicator that is used in many circuits to-

day. Fig. 10-19 shows the schematic symbol for a LED.

The light produced by a LED may be one of several colors: red, green, yellow or amber. LEDs are very small in size compared to incandescent (filament) lamps. They use very little power and have an extremely long life. Also, they give off light much quicker than a filament lamp.

Fig. 10-18. A typical silicon diode used in rectifier circuits. It is nicknamed the "top hat" because of its shape. (International Rectifier Corp.)

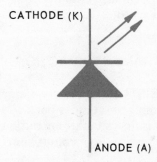

Fig. 10-19. Symbol for Light Emitting Diode (LED)

LEDs are made of either gallium arsenide or gallium arsenide phosphide. These materials give off light when current is passed through them in the right direction.

LED devices must be operated within the current limitations listed in the publications of the manufacturer. Peak current conditions, also listed, must never be exceeded.

When soldering LED devices, use standard transistor and IC techniques. That is, place a heat sink on the lead being soldered. Place it between the device and the point to which the lead is being soldered.

Light emitting diodes may come in single indicator packages, Fig. 10-20, or they may

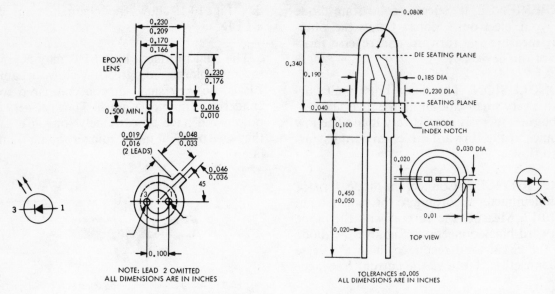

Fig. 10-20. Two package styles of LEDs.   (Sprague Products Co.)

come in 7-segment alpha-numeric display found on many calculators, clocks and watches. See Fig. 10-21.

TRANSISTOR ACTION

A TRANSISTOR is a three-element device made by forming two diodes back to back. The interesting results should encourage you to study this fascinating component in later courses. In Figs. 10-22 and 10-23 are block diagrams with a lot of information for an NPN transistor and again for the PNP. Two batteries are used, at this time, to assist you in understanding the action.

Note that in each case, the NPN and the PNP

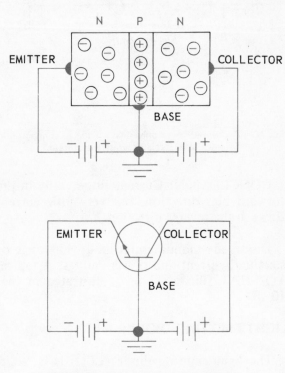

Fig. 10-21. Alpha-numeric LED Display.
(Sprague Products Co.)

Fig. 10-22. NPN transistor connections.

128

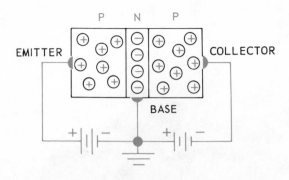

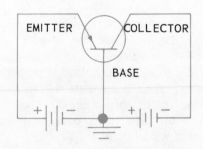

Fig. 10-23. PNP transistor connections.

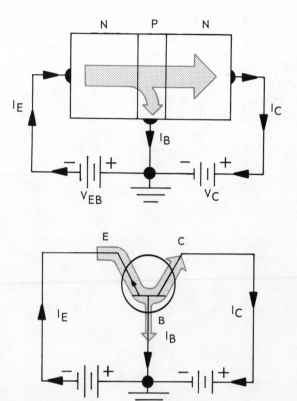

Fig. 10-24. Only a small percentage of the current flows in the base circuit.

transistor are connected so that the EMITTER-BASE diode is FORWARD BIASED. This is the low resistance direction.

Note, too, that the COLLECTOR-BASE diode is REVERSE BIASED in each type of transistor. This is the very high resistance direction.

TRANSISTOR CURRENTS

Study the diagram in Fig. 10-24. Because the emitter-base circuit is forward biased, a current flows into the EMITTER (E). The BASE (B) section is very thin, and electrons can quite easily cross the base-collector junction. As a result, about 99 percent or more of the emitter current flows directly to the COLLECTOR (C). Only one percent flows in the BASE CIRCUIT. These simple equations are true:

$$I_E = I_B + I_C$$

$$I_C = I_E - I_B$$

CURRENT GAIN

The output current I_C must be LESS THAN

the input current I_E. There can be no gain in current. In fact, it must be less than one. The current gain (less than one) is called ALPHA ($\propto$). It is equal to: $\dfrac{I_C}{I_E}$.

Assume there is an emitter current of 10 mA and a base current of .1 mA. Then:
$$I_C = I_E - I_B = 10 \text{ mA} - .1 \text{ mA} = 9.9 \text{ mA}$$

$$\text{Current gain} = \frac{I_C}{I_E} = \frac{9.9}{10} = .99 = \propto$$

VOLTAGE GAIN

Voltage gain is a different story. The resistance of the forward biased EB junction is low, and the resistance of the CB junction is very high. About the same current flows through both junctions. Therefore, the voltage across the EB junction is low and the voltage across the CB junction is high ($I \times R = V$).

A circuit similar to a transistor shows these resistances and currents in Fig. 10-25. The voltage "in" and across junction EB is $I_E \times R_{EB}$. The voltage "out" is $I_C \times R_{CB}$. Using our previous current figures, and assuming $R_{EB} = 50 \ \Omega$ and $R_{CB} = 500,000 \ \Omega$, then the voltage gain (A_v) of the transistor is:

$$\frac{\text{OUTPUT}}{\text{INPUT}} \quad \frac{9.9 \text{ mA} \times 500,000 \ \Omega}{10 \text{ mA} \times 50 \ \Omega}$$

$$= \frac{4950V}{.5V} = 9900$$

This gain is reduced by other resistors and components in a circuit, but it still is very high.

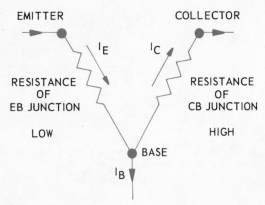

Fig. 10-25. I_E is flowing through low resistance. V_E is a small voltage. I_C (which is only 1 percent less than I_E) flows through high resistance R_{CB} and produces a high voltage.

The previous discussion tells you how the TRANSISTOR got its name. TRANS stands for transfer and "sistor" for resistance. A transistor is a device which realizes its gain by a "transfer of resistance" or "transistor." A typical transistor tester is shown in Fig. 10-26.

THE COMMON EMITTER CIRCUIT

By far the most popular way of connecting a transistor in a circuit is the COMMON EMIT-TER or CE method drawn schematically in Fig. 10-27. The input signal is applied between the base and emitter; the output is taken from the collector and emitter. The emitter is common to both circuits. A small voltage applied to the input produces a relatively large voltage across R_C in the output.

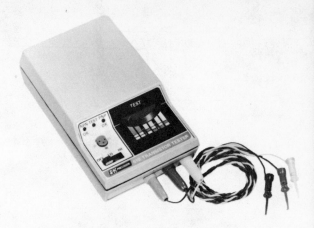

Fig. 10-26. Transistor tester.

R_C is the load resistor, and it works exactly the same as when used with the vacuum tube studied earlier in this chapter. The common emitter circuit also turns over or inverts the signal like a vacuum tube.

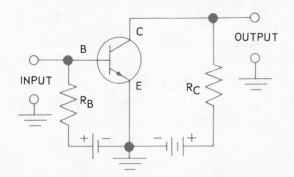

Fig. 10-27. COMMON-EMITTER circuit connection. The emitter is grounded and common to both the input and output circuits.

COMMON EMITTER CURRENT GAIN

These examples show the ability of the common emitter circuit to amplify current.

Assume as before, the emitter current of some transistor is 10 mA and the collector current is 9.9 mA. Then we can say that 99 percent of the total current flows in the collector circuit and only one percent in the base circuit. This percentage ratio stays more or less constant under a given set of circuit conditions. In other words, if the emitter current is increased to 20 mA, then the collector current increases to 99 percent of 20 mA or 19.8 mA and the base current is .2 mA.

We can say then that a small change in BASE CURRENT results in a large change in COLLECTOR CURRENT. This relationship is called the BETA (β) of a transistor. It can be stated mathematically as:

$$\beta_{dc} = \frac{I_C}{I_B}$$

In our example:

$$\beta_{dc} = \frac{9.9 \text{ mA}}{.1 \text{ mA}} = 99$$

CASCADED TRANSISTORS

We talk about waterfalls on a mountain stream as CASCADES. One falls above another. To increase the gain or amplification of a circuit, several transistors are coupled together in cascade fashion. The output of one feeds the input of the next stage.

SIREN PROJECT

The siren shown in Fig. 10-28 produces a quickly rising "wail" when the push button switch is depressed on the front panel. After the push button switch is released, the siren sound coasts down. The sound produced by this novel project resembles that produced by a motor-driven siren.

Fig. 10-28. The Siren with push button switch.

The schematic for the siren is shown in Fig. 10-29. This project uses a minimum of component parts. Note that a higher impedance, 45 ohm speaker is required.

FORWARD STEPS IN UNDERSTANDING ELECTRONS

1. The Edison Effect is the formation of a space cloud of electrons around a heated cathode.
2. A diode has a cathode and a plate. It conducts in only one direction.
3. A diode is used in rectification to change ac to dc.
4. A triode has a cathode, plate and grid.
5. The grid in a triode controls the electron current flowing through the tube.
6. Amplification results when a small change in input voltage causes a large change in output voltage.
7. A triode amplifier inverts the signal.
8. Bias is the fixed voltage applied to the grid of a tube. Usually, it is negative.
9. A tube can act as a switch by switching grid voltage from conduction to cutoff.
10. A P crystal has HOLES for current carriers.
11. An N crystal has electrons for current carriers.
12. A solid state diode conducts when forward biased.
13. A diode has high resistance in the reverse direction.
14. A transistor is a three element device and can be either NPN or PNP.
15. The current gain of a transistor connected as a common base circuit is called "alpha."
16. $I_E = I_C + I_B$, $I_C = I_E - I_B$,

   $$\propto = \frac{I_C}{I_E}, \qquad I_C = \propto I_E$$

17. The common emitter is the most popular transistor connection. It produces both current gain and voltage gain.
18. Beta is the current gain of a common emitter circuit, and it indicates the ability of the circuit to amplify.
19. Transistors can be coupled in cascade for greater amplification.

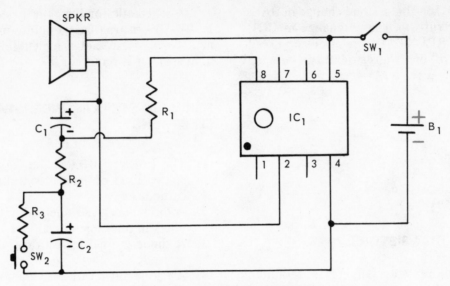

PARTS LIST FOR THE SIREN

R_1 — 200 Ω, 1/2W resistor
R_2 — 2.7 KΩ, 1/2W resistor
R_3 — 470 Ω, 1/2W resistor
C_1 — electrolytic capacitor, 1 μF @ 6 WVdc
C_2 — electrolytic capacitor, 500 μF @ 3 WVdc
IC_1 — integrated circuit unit, National

Semiconductor LM 3909
SW_1 — SPST switch
SW_2 — push button switch, N.O.
SPKR — 25-45 Ω speaker
B_1 — 1 1/2V "D" cell
Misc. — case, PC board and materials, IC socket, wire, solder, decals

Fig. 10-29. Schematic and parts list for the Siren.　(National Semiconductor Corp.)

RULES OF SAFETY

SAVE THE TRANSISTORS

Before making changes in a transistor circuit, always disconnect the power. This is also required if you remove a transistor from its socket. Voltage surges can destroy transistors.

TRANSISTORS CAN BE DESTROYED

When connected correctly, transistors are rugged little components. But if connected backwards with wrong voltage polarity, they can be destroyed. Check your connections BEFORE applying the power.

TEST YOUR KNOWLEDGE - UNIT 10

1. Build a TWO TRANSISTOR AMPLIFIER CIRCUIT. Consult the diagram in Fig.

10-30 and the suggested layout in Fig. 10-31. Adjust voltage to 15 volts and turn the amplifier on.

A. Tap the microphone with your finger. The tap should be heard from the speaker. A very small voltage produced by the microphone is now amplified.

B. Place microphone near speaker. Does it squeal? This is caused by feedback. The noise from the speaker enters the microphone and is reamplified and around and around it goes.

C. Remove C_2 from the circuit. Does the amplifier work better or worse? The removal of C_2 reduces the gain of the amplifier by degeneration (feedback 180 deg. out-of-phase with input signal, so it subtracts from input). The voltage at the emitter varies in such a manner as to oppose the input.

D. Connect record player to your amplifier. Does it amplify with reasonable fidelity?

E. Team up with a friend who has built a similar amplifier and build an intercom. You will need longer leads.

F. Connect an audio generator (see page 187) to the input of your amplifier. Set the frequency at about 400 Hz. Connect your oscilloscope across the input and adjust to display four waves about .2 volt from peak-to-peak. Connect the oscilloscope across the speaker terminals and observe the wave.

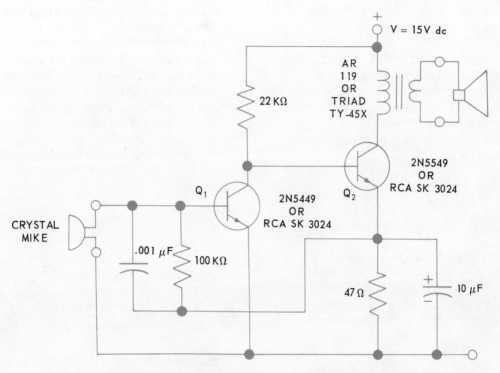

Fig. 10-30. Schematic diagram for the experimental amplifier.

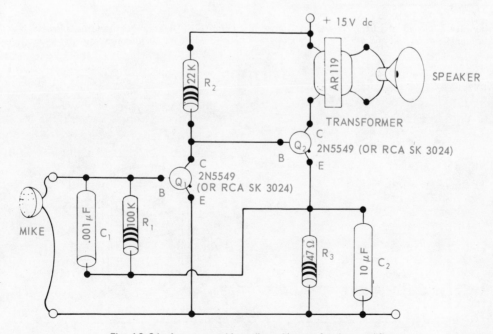

Fig. 10-31. A suggested breadboard layout for the amplifier.

Can you figure out the voltage gain? Is the output wave distorted? If so, reduce the AF (audio frequency) generator output until you have an undistorted wave.

2. Construct a TRANSISTOR SWITCH. To show the ability of a transistor to act as a switch, build the circuit in Fig. 10-32. Explain the transistor action.

3. Build a TRANSISTOR RELAY. In order to switch higher currents and power circuits, the relay circuit is constructed as in Fig. 10-33. The transistor circuit operates on 15 volts dc. The lamp circuit uses 6.3 volts ac. Explain the transistor action.

4. What are the advantages of using transistors over vacuum tubes?

5. In an N crystal, the current carriers are_____ _____.

6. In a P crystal, the current carriers are_____ _____.

7. Two transistors are connected in cascade. One has a voltage gain of 50 and the other 60. What is the total voltage gain?

8. Build a SALINITY TESTER to test "how salty?" Water decreases its resistance when salt is added. In the circuit of Fig. 10-34, a glass jar filled with water is used. The test probes are pieces of bare copper wire. Add salt to the water a little at a time. Watch the current meter. The meter could read in percentage of salt rather than mA.

Explain the transistor action.

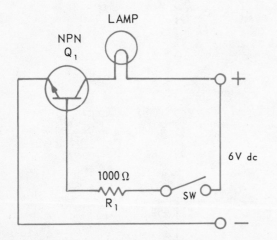

Fig. 10-32. When SW is closed, the lamp glows. The transistor is turned "on and off."

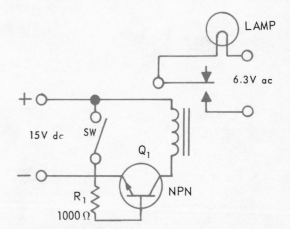

Fig. 10-33. A relay activated by a transistor switch. The relay contacts could control a high powered machine.

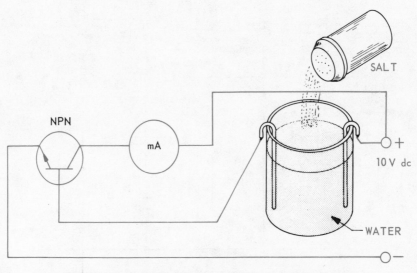

Fig. 10-34. The salinity of the water will be indicated on the meter.

Unit 11

INTEGRATED CIRCUITS

Development of the INTEGRATED CIRCUIT was one of the most important inventions in the field of electronics. Major topics discussed in this Unit include:

1. What active and passive devices are.
2. How digital systems differ from linear systems.
3. How integrated circuits are made.
4. What symbols and outlines are used for integrated circuits.
5. What the binary numbering system is.
6. How the basic gates used in computers operate.
7. What the basic block diagram of the computer is.

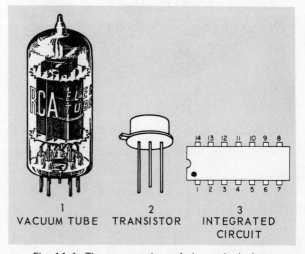

Fig. 11-1. Three generations of electronic devices.

INTEGRATED CIRCUITS

Integrated circuits (ICs) are the third generation of major electronic devices. The first semiconductor (crystal) diode was developed in 1883. In 1904, the first vacuum tube diode (called the "valve") was introduced. Just a couple of years later, the triode vacuum tube capable of amplifying was developed by Dr. DeForest (See Unit 10). VACUUM TUBES are considered as the first generation of electronics devices.

In 1948, the TRANSISTOR was invented, and it became the second generation of electronics devices. This was a very important achievement in the electronics field, since the transistor has many advantages over the vacuum tube.

The third generation of electronic devices is the INTEGRATED CIRCUIT. See Fig. 11-1.

ICs were developed in 1958, only ten years after the invention of the transistor. An integrated circuit is made of a specially processed chip of silicon to form hundreds or thousands of transistors, resistors, capacitors and similar components. Fig. 11-2 compares the size of a transistor to many IC chips.

Once, it was the practice to manufacture separate or discrete components, then assemble them together by wires in a chassis or on a printed circuit. In recent years, with the development of integrated circuits and microelectronic circuits, this has changed. Transistors, diodes, resistors, capacitors and other components and their interconnections are now assembled on to one single silicon chip smaller than a lower case "o" on this printed page. It is now possible to package thousands of components (or elements) on one small IC chip.

135

Fig. 11-2. A comparison of the size of a transistor with several hundred integrated circuit (IC) chips.
(U.S. Atomic Energy Commission)

Fig. 11-3 shows a microminiature television set which uses ICs.

Compared to transistors, ICs have many advantages and some disadvantages. See Fig. 11-4.

ACTIVE VS PASSIVE DEVICES

An ACTIVE DEVICE used in electronic circuits has the ability to either change its state or control one type of signal with another one. Ac-

Fig. 11-3. A microminiature television set.
(Sinclair Electronics)

INTEGRATED CIRCUITS

ADVANTAGES	DISADVANTAGES
1. Low cost 2. Higher switching speed 3. Low power consumption 4. Small size	1. Certain components cannot be fabricated in ICs. 2. Cannot handle large amounts of voltage or current

Fig. 11-4. Advantages and disadvantages of ICs when compared with transistors.

tive devices, such as transistors and integrated circuits, can change their state in response to some outside signal.

A PASSIVE DEVICE does not have this control ability, nor does it have any gain or amplification. Examples of passive devices are resistors, capacitors and inductors (coils).

An integrated circuit can be fabricated to have both active and most passive components. Semiconductors such as transistors and diodes can be fabricated into a chip of silicon. Limited values of resistance and capacitance can also be built into a silicon material. Inductors or coils, however, cannot be fabricated into an IC and must be placed as a separate component in the circuit.

BASIC ELECTRONICS SYSTEMS THAT USE INTEGRATED CIRCUITS

Electronic systems that use ICs usually are classified as either digital or linear (analog). DIGITAL systems work on or manipulate data in separate bits or pieces. Usually, these bits or pieces are conditions in a circuit where the current flows or does not flow, or a magnet is magnetized in one direction or the other direction. See Fig. 11-5.

LINEAR electronic systems operate on data or information that is continuous over a range or variable. Sometimes linear is referred to as "Analog." Linear integrated circuits control electricity over a wide range, Fig. 11-5, rather than switching it suddenly between certain levels.

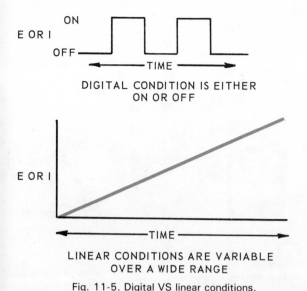

DIGITAL CONDITION IS EITHER
ON OR OFF

LINEAR CONDITIONS ARE VARIABLE
OVER A WIDE RANGE

Fig. 11-5. Digital VS linear conditions.

HOW INTEGRATED CIRCUITS ARE MADE

To understand the industrial process of manufacturing integrated circuits, refer to Fig. 11-6. A detailed explanation of each of these steps follows:

1. The CIRCUIT DESIGNER plans the total integrated circuit, which includes the purpose of the chip and the schematic diagram.
2. From the schematic diagram produced by the circuit designer, the LAYOUT DESIGNER plans the actual, large scale design of the IC. Many ICs are laid out with the aid of a computer. This layout is often checked by using an enlarged image of the integrated circuit, which is 500 times larger than the actual size of the IC device to be manufactured.
3. The next step is the PLANAR COORDINATOGRAPH. When the design and layout (steps 1 and 2) of a new circuit are complete, a computer is used to "memorize" the location of the actual positions of each element in the circuit so that they fit accurately on top of each other.
4. Now, each layout of the circuit is PHOTOGRAPHICALLY REDUCED to a microminiature size. The concept of this process is: the fabrication of many circuits at a time; the reducing the circuits to the smallest possible size; and the maximum

simplification of the processing technology.

5. Next, a set of working plates is made by the process of CONTACT PRINTING. These plates are used in the photomasks.
6. The PHOTOMASKS are made from computer memorized data of the exact location of each element in an IC. Each mask holds the "patterns" of each single layer of the circuit.
7. The next step is to GROW THE CRYSTAL. Crystals are made from raw silicon, which is first reduced from its oxide (the main makeup of common sand). The purification process used in industry can refine silicon to 99.9999999 percent pure (one impure part in 1,000,000 pure parts of silicon). Then, the manufacturer adds the needed impurities to make N type or P type areas in the silicon. This process of adding the impurities is called DOPING. The term "growing" comes from withdrawing a small doped silicon "seed" from a quartz crucible containing molten silicon to form a crystal three to four inches in diameter. This may be compared to inserting a stick into melted caramel and pulling up the stick and some caramel from the container.
8. The grown crystal is then SLICED into a wafer and POLISHED. The slicing process is accomplished by cutting them into wafers about the thickness of a postcard (about .5 mm thick). The polishing process removes any surface scratches.
9. From the thin wafers of doped silicon, the BASIC BUILDING BLOCK process begins by forming electronic patterns on the surface of the silicon. The microelectronic circuit is built up layer by layer on the silicon wafer, each layer receiving a pattern from the photomask used in the circuit design.
10. Next, the DIFFUSION and REOXIDATION process takes place in the manufacturing of integrated circuits:
   a. First, an epitaxial of N type silicon about .25 microns thick is grown in the wafer. NOTE: Epitaxy is the physical placement of materials on a surface. This layer ultimately becomes the collector for transistors or an element of a diode or capacitor.

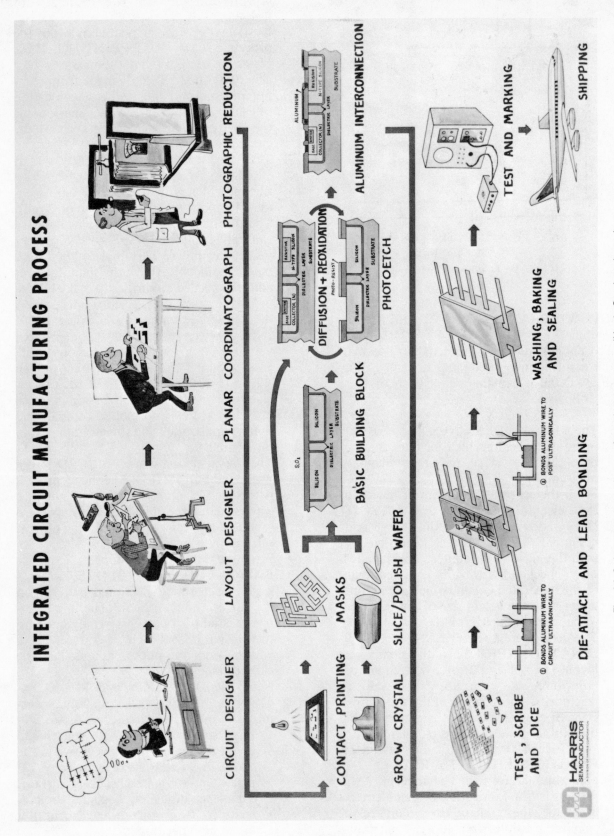

Fig. 11-6. Integrated circuit manufacturing process.   (Harris Semiconductor)

b. Next, a thin coat of silicon dioxide is grown over the N type material by exposing the wafer to an oxygen atmosphere at about 1000°C (1832°F). Steps 1 and 2 are illustrated in Fig. 11-7.

c. To prepare the wafer for isolation between various components, the wafer is covered with photo-resist and exposed under ultraviolet light through a specific photographic mask. The nature of the mask depends upon the circuits to be made.

d. Then, the wafer is etched with hydrofluoric acid and unexposed areas of silicon dioxide are etched away. Steps 3 and 4 are shown in Fig. 11-8.

e. Next, the wafer is subjected to a diffusion process using boron. The boron diffuses into and forms a P type material on all areas not protected by the silicon dioxide. Sufficient time is allowed for DIFFUSION completely through the epitaxial layer to the P type substrate. The wafer now appears as in Fig. 11-9 with isolated islands of N type material. Isolation is realized by the formation of the NP junctions around each island, and there are back-to-back diodes between each N type island.

f. During diffusion, a new layer of silicon dioxide forms over the diffused P type areas as well as on the top of the islands.

g. Using the photo-resist coating again and exposure under a specified mask, areas in the N type islands are etched away. Once again, the wafer is subjected to a P type diffusant and areas are formed for transistor base regions, resistors or elements of diodes or capacitors.

h. Again, the wafer is REOXIDIZED. Refer to Fig. 11-10.

i. The wafer is again masked, exposed and PHOTO-ETCHED to open windows in the P type regions. Next in the IC manufacturing process, a phosphorus diffusant is used to produce N type regions for diodes and capacitors. Small windows are also etched through to the N layer for electrical connections. The total wafer is again given the oxide coating. See Fig. 11-11.

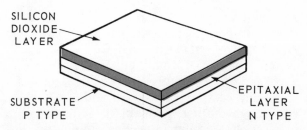

Fig. 11-7. Steps 1 and 2 in the process of making an IC.

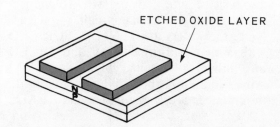

Fig. 11-8. The first masking and etching is for isolation of components.

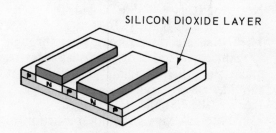

Fig. 11-9. N type material islands remain after P diffusion.

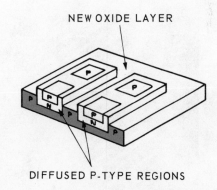

Fig. 11-10. P type regions are diffused in the N type islands.

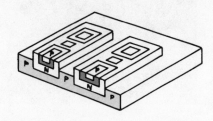

Fig. 11-11. Emitters are diffused into the P type regions.

11. For the particular IC, the monolithic circuit is complete except for the ALUMINUM INTERCONNECTIONS. A thin coating of aluminum is vacuum-deposited over the entire circuit. Then, the aluminum coating is sensitized and exposed through another special mask. After etching, only the interconnecting aluminum forms a pattern between transistors, diodes and resistors and pads for wires to connect the wafer to an internal circuit.

12. Next, the wafers are TESTED, SCRIBED with a diamond-tipped tool and DICED into individual circuits.

13. After separation, the individual circuits are mounted on a ceramic wafer and LEADS (.001 in. or 0.025 mm diameter) are BONDED. The circuit may be mounted in a small can or flat package. An IC with leads bonded is shown in Fig. 11-12.

14. Next, the IC is WASHED, and, the cavity holding the IC is SEALED.

15. Then, the integrated circuit goes through an exhaustive series of electrical tasks to make sure that the unit performs as it should.

16. Lastly, the circuit is shipped to the distributor. A very dramatic reduction in the costs of integrated circuits have taken place in the past few years. Along with the overall dropping of prices of ICs, the quality of each unit has improved greatly. Fig. 11-13 shows a high density integrated circuit that is used as a single board computer chip.

SYMBOLS FOR INTEGRATED CIRCUITS

Usually, integrated circuit symbols used in schematic diagrams are shown in a rectangular or triangular shape. Fig. 11-14 shows these two styles of symbols.

OUTLINES FOR INTEGRATED CIRCUITS

Types of basic IC designs along with their pin numbering systems and dimensions are shown in Fig. 11-15.

BASIC COMPUTER LOGIC

In a modern computer, the ON and OFF operation is not performed by mechanical switches since that would be much too slow. Instead, this operation is accomplished by integrated circuits. However, the theory of switching is the same.

Fig. 11-12. An integrated circuit with leads bonded. (U.S. Atomic Energy Commission)

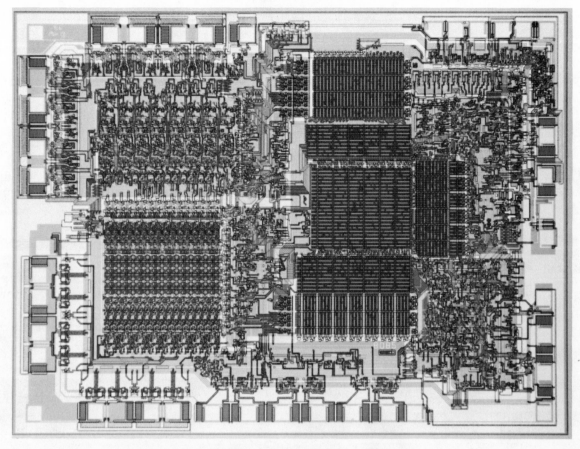

Fig. 11-13. Single board computer chip. (Intel Corp.)

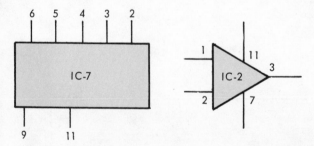

Fig. 11-14. IC symbols. Note pin numbering system.

In computers, a switch is considered to be in one of two states, either ON or OFF. This relates to BINARY (base two) numbers in which ON is represented by number 1 and OFF by 0. You may also read your own answers to any specific questions. Use YES for 1 or switch ON. Use NO for 0 or switch OFF.

With this in mind, look at the decision circuit of two switches in Fig. 11-16. The circuit shown is known as an AND gate. This is another way

of saying: IF SW$_1$ IS YES OR ON AND SW$_2$ IS YES OR ON, THEN THE CIRCUIT IS COMPLETE AND THE LAMP GLOWS IN A POSITIVE YES. Also note the symbol for the AND gate used in computer diagrams and the typical TRUTH TABLE showing the four possible conditions of the circuit.

THE OR GATE CIRCUIT

The arrangement of another popular computer circuit is illustrated in similar fashion in Fig. 11-17. In this case, SW$_1$ and SW$_2$ are in parallel and it can be stated that: IF EITHER SW$_1$ OR SW$_2$ IS CLOSED, THERE WILL BE AN OUTPUT AND THE LAMP WILL GLOW. TRUTH TABLE for this circuit is also shown in Fig. 11-17.

Simple and very complex decisions can be made by using the circuits alone and in combination.

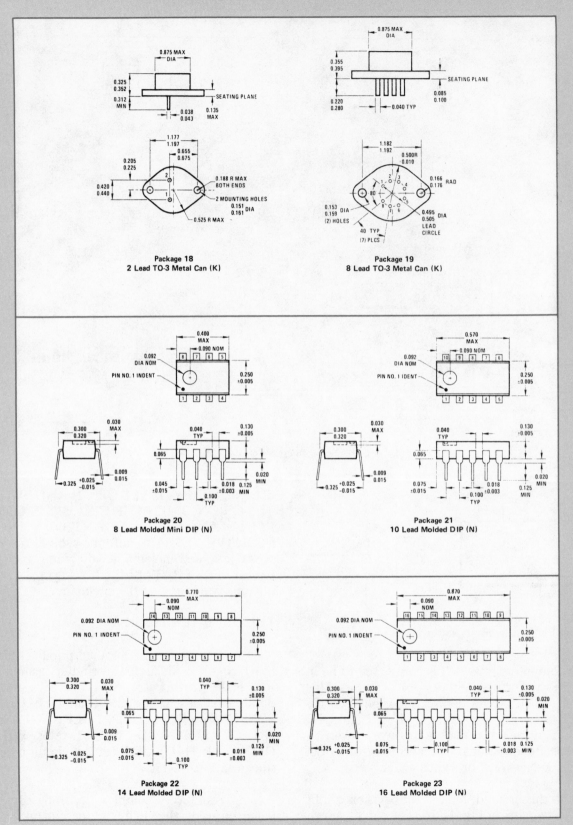

Fig. 11-15. Outlines for ICs.   (National Semiconductor)

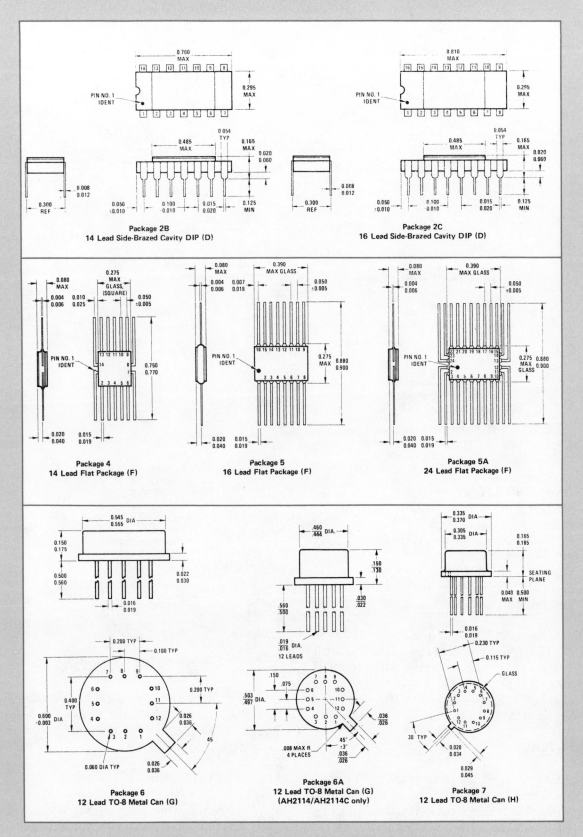

Fig. 11-15. (Continued)   Outlines for ICs.   (National Semiconductor)

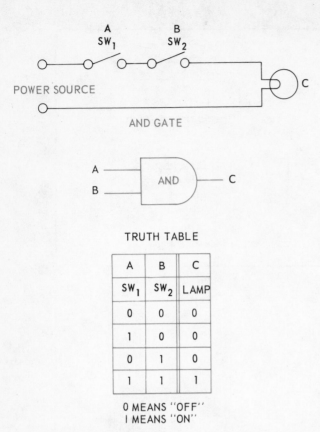

TRUTH TABLE

A	B	C
SW$_1$	SW$_2$	LAMP
0	0	0
1	0	0
0	1	0
1	1	1

0 MEANS "OFF"
1 MEANS "ON"

Fig. 11-16. AND GATE SWITCHING. Note that both SW$_1$ AND SW$_2$ must be in the ON or 1 position to realize an output 1 indicated by glowing lamp.

A COMPUTER STORY

Suppose you want to go to the basketball game this Friday night, but your parents have set up these conditions:
A. You must do your homework.
B. You must find a friend to go with you.
C. You must promise to be home at 10 o'clock.

This is written by using logic symbols as in Fig. 11-18. If the three inputs A, B and C are 1 or YES, then D is YES with permission to go. If any one of the conditions is NO or 0, then D will also be NO. You will have to stay home and watch the game on TV.

INVERTERS

An inverter is added to the basic gate circuit to produce an answer opposite to the original. The symbol is a small circle illustrated in Fig. 11-19 for both the AND and OR gates. With the inverter added the gates become NAND (not

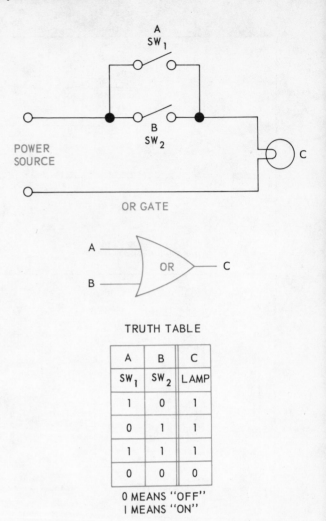

TRUTH TABLE

A	B	C
SW$_1$	SW$_2$	LAMP
1	0	1
0	1	1
1	1	1
0	0	0

0 MEANS "OFF"
1 MEANS "ON"

Fig. 11-17. OR GATE SWITCHING. Note the lamp will glow showing a complete circuit if either SW$_1$ or SW$_2$ is closed or in 1 position.

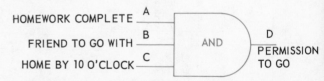

Fig. 11-18. Logic symbol for computer story in text.

and) and NOR (not or). For the NAND gate, ONES at A and B produce ZERO for output. For the NOR gate, ONES at either A or B produce ZERO for output.

LOGIC EQUATIONS

When writing the description of gates and logic diagrams, the correct expressions must be used.

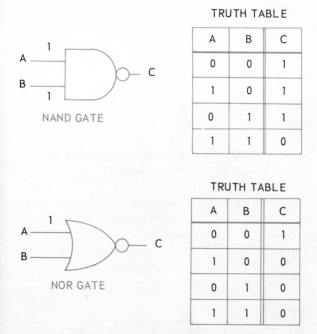

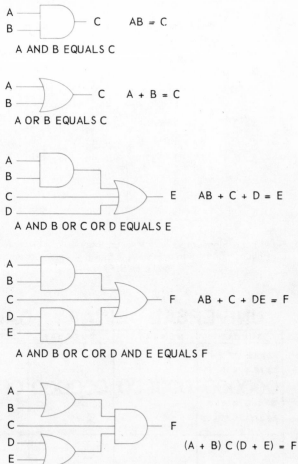

Fig. 11-19. Symbols and truth tables for NAND and NOR gates.

For the AND gate. An input at A and B produces output C. This is written:

$$AB = C$$

For the OR gate. An input at A or B produces output C. This is written:

$$A + B = C$$

Study the symbols and equations of the logic diagrams in Fig. 11-20. Write a TRUTH TABLE for each equation.

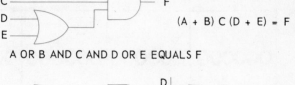

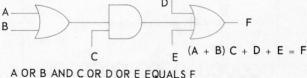

Fig. 11-20. Study the logic symbols and equations.

COUNTING CIRCUITS

We are used to counting by the digit system, and we use numbers from 0 to 9. In the BINARY system, counting is possible using only 0 and 1. Any of the familiar numbers can be represented by some combination of ONES and ZEROS. In computers, the ONES and ZEROS can represent different voltage levels on a line or wire. For example, we might consider zero voltage as binary number 0 and a positive 10 volts as number 1.

Consider the chart in Fig. 11-21 in which only three wires are used. By changing the voltage from 0 to 10 or the signal from 0 to 1 on these wires, they can represent any number between 0 and 7. By adding a fourth wire, we could count to 15 and so on. Study the makeup of the chart in Fig. 11-21.

Fig. 11-22 presents a design for a binary calendar. This universal page can be duplicated and filled in for each month of the year. Start by making the first day of each month read 000000; mark the second day 000000; the third day is 000000, etc.

FLIP-FLOPS

A very useful circuit in the computer is the FLIP-FLOP, which usually consists of two transistors or an integrated circuit. When one is conducting or ON, the other is OFF. A pulse of voltage will cause the circuit to flip, and number one will turn OFF and the second will turn ON. Another pulse will cause the circuit to flop back to its original state. Indicating lights are used to show which transistor is conducting. The light ON represents a binary 1, the light OFF is a binary 0. In Fig. 11-23, four of these flip-flops (note the logic symbol) are connected in CASCADE. Now examine what happens when pulses are applied to the circuit input as illustrated in Fig. 11-24.

WIRE NUMBER

DIGIT NUMBER		3	2	1	BINARY NUMBERS
0		0	0	0	000
1		0	0	1	001
2		0	1	0	010
3		0	1	1	011
4		1	0	0	100
5		1	0	1	101
6		1	1	0	110
7		1	1	1	111
8	1	0	0	0	1000

Fig. 11-21. Signal or voltage level on three wires can represent binary numbers from 0 to 7. A fourth wire is needed to make an 8.

UNIVERSAL BINARY CALENDAR

SUNDAY	MONDAY	TUESDAY	WEDNESDAY	THURSDAY	FRIDAY	SATURDAY
POWER OF 2: 32 16 8 4 2 1 OOOOOO 5 4 3 2 1 0 POSITION VALUE	OOOOOO	OOOOOO	OOOOOO	OOOOOO	OOOOOO	OOOOOO
OOOOOO	OOOOOO	OOOOOO	OOOOOO	OOOOOO	OOOOOO	OOOOOO
OOOOOO	OOOOOO	OOOOOO	OOOOOO	OOOOOO	OOOOOO	OOOOOO
OOOOOO	OOOOOO	OOOOOO	OOOOOO	OOOOOO	OOOOOO	OOOOOO
OOOOOO	OOOOOO	OOOOOO	OOOOOO	OOOOOO	OOOOOO	OOOOOO

BOB STOVER, STUDENT AT HAMMOND HIGH SCHOOL — ALEXANDRIA, VIRGINIA

Fig. 11-22. A universal binary calendar.

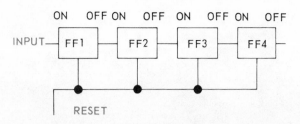

Fig. 11-23. Four flip-flops connected in cascade. Each has an indicating light.

INTEGRATED CIRCUITS AND MODERN COMPUTERS

With modern day technology, it is now possible to fabricate complete computing systems on a single integrated circuit chip. See Fig. 11-25. The actual size of this IC is 5.6 × 6.6 mm (millimetres) or approximately the size of the flat surface of a pencil eraser. This microcomputer will hold 8,192 bits of information in the memory. Also, the central processing unit (CPU) is built into the IC chip.

The basic block diagram for a microcomputer is shown in Fig. 11-26.

The INPUT/OUTPUT part of the computer provides a means of communicating with the inside of the computer. Instructions and data are fed in through input devices such as a keyboard, Fig. 11-27, cassette tapes, floppy discs, Fig. 11-28, paper tape or punched cards. These devices also act as outputs so the computer can communicate with the user. Exterior units such as video displays are often called "peripherals" in computer language.

The MEMORY shown in the block diagram in Fig. 11-26 stores data and programs. This storage usually is in binary words ("0s" usually are low voltage pulses and "1s" are high voltage

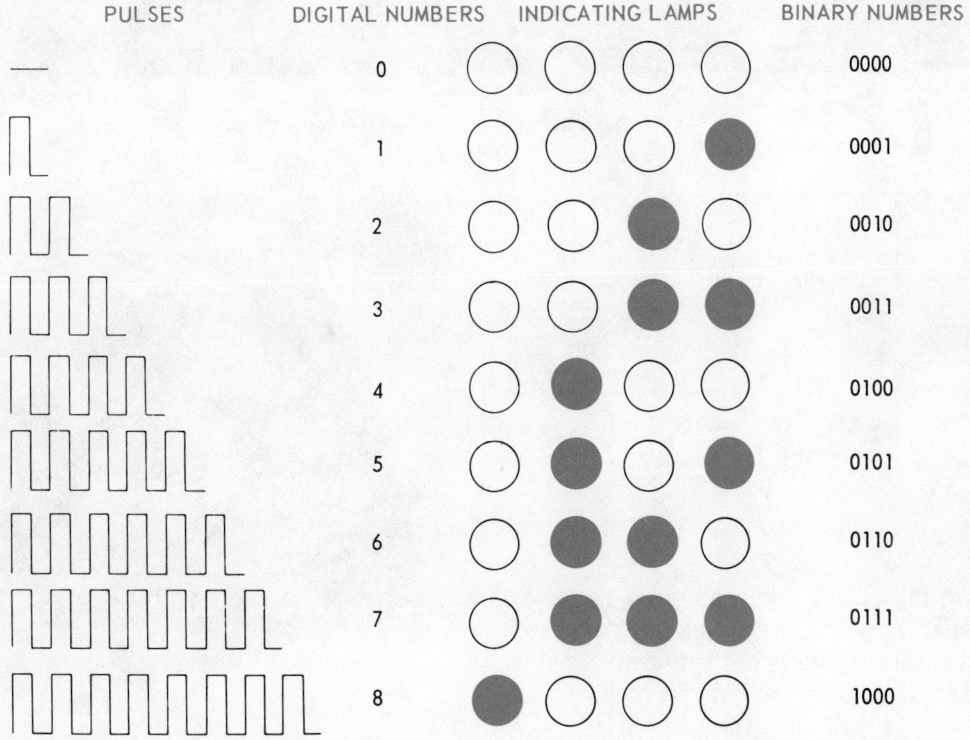

PULSES	DIGITAL NUMBERS	INDICATING LAMPS	BINARY NUMBERS
	0	○ ○ ○ ○	0000
	1	○ ○ ○ ●	0001
	2	○ ○ ● ○	0010
	3	○ ○ ● ●	0011
	4	○ ● ○ ○	0100
	5	○ ● ○ ●	0101
	6	○ ● ● ○	0110
	7	○ ● ● ●	0111
	8	● ○ ○ ○	1000

Fig. 11-24. The indicator lights show the binary number corresponding to input pulses.

Fig. 11-25. Single chip microcomputer. (Intel Corp.)

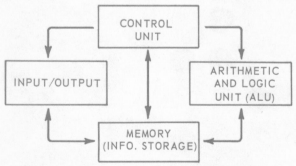

Fig. 11-26. Basic block diagram of a computer.

pulses). See Fig. 11-29 for a memory module. The CONTROL UNIT interprets or "decodes" the program and produces the necessary signals to make the arithmetic unit perform its function. The ARITHMETIC AND LOGIC UNIT (ALU) performs the arithmetic and logic functions on the data such as adding or shifting.

Fig. 11-27. Computer input keyboard. (Processor Technology)

Fig. 11-28. Computer input floppy disc.   (Processor Technology)

Fig. 11-29. Computer memory module used in a microcomputer.   (Processor Technology)

A complete microcomputer system showing the input/output terminal, video display terminal and a dual floppy disc storage unit is shown in Fig. 11-30. This type of microcomputer system can be used in small businesses, classrooms and homes. Some specific home uses for these personal computers are to help with updating checking accounts; to tutor students with math; to entertain the family with computer games; to help plan menus; to convert measurements; to help in conserving household energy consumption.

BINARY TO DECIMAL COUNTER PROJECT

The binary to decimal counter in Fig. 11-31

Fig. 11-30. A complete microcomputer system used in small businesses and homes.   (Processor Technology)

will take pulses fed into a push button switch and change them into digit counts that can be read on a LED display. After the counter reaches 9, it will return to 0 and continue counting. This basic circuit is fundamental to all counting circuits in electronic units such as calculators and digital voltmeters.

A commercial single digit counter board is used in this project. See Fig. 11-32 for a schematic and parts list. Assemble the printed circuit board from the instructions from Radio Shack. Then, experiment with the LED display to see how you can light the various elements.

FORWARD STEPS IN UNDERSTANDING ELECTRONICS

1. Integrated circuits are the third generation in the family of electronic devices.
2. An integrated circuit (IC) is a specially processed chip of silicon to form hundreds or

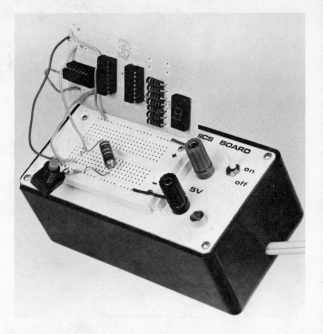

Fig. 11-31. Binary to Decimal Counter.

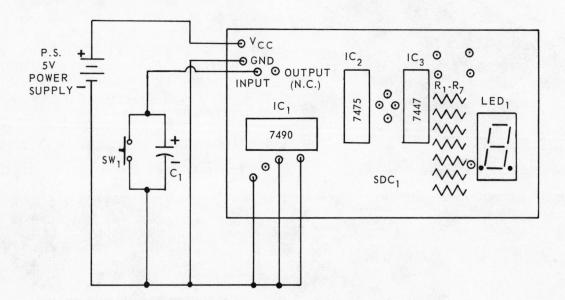

PARTS LIST FOR BINARY TO DECIMAL COUNTER

C_1 — electrolytic capacitor, 1 μF @ 3 WVdc
SW_1 — push button switch, N.O.
PS — 5V power supply
SDC_1 — single digit counter project board (Radio Shack #277-103) or the following components:

IC_1 — integrated circuit #7490
IC_2 — integrated circuit #7475
IC_3 — integrated circuit #7447
LED_1 — .3 in. light emitting diode, single digit readout
R_1-R_7 — 150 Ω, 1/2W resistors
Misc. — case, wire, IC sockets, solder, decals

Fig. 11-32. Parts list and schematic for Binary to Decimal Counter.

thousands of transistors, diodes, resistors and capacitors.

3. An active electronic device is one which has the ability to change its state or control one type of signal with another signal.

4. A passive electronic device does not have the ability to control, only to block or restrict current or voltage.

5. Integrated circuits can be fabricated to have active and passive devices.

6. Integrated circuits can be broadly classified as digital or linear.

7. Symbols used for ICs are made in rectangular or triangular form.

8. There are a number of different types of packaging outlines for ICs.

9. The binary numbering system used in many computers is based on two digits or states: 0 and 1.

10. The basic block diagram for a computer includes: an input/output; a control unit; a memory or information storage; an arithmetic and logic unit.

RULES OF SAFETY

CAUTIONS IN USING INTEGRATED CIRCUITS

ICs are heat sensitive so use a socket, if possible, in mounting the device into the circuit. Do not solder ICs into a circuit with a high wattage soldering gun.

ICs require careful handling because of their small size and many contacts. Certain units have as many as 12 flexible leads spaced around a circle less than 1/4 in. in diameter. This arrangement calls for patient handling to avoid bends and short circuits.

TEST YOUR KNOWLEDGE - UNIT 11

1. _____ are considered to be the first generation electronic devices.

2. Which element is the basic material used to make ICs?

3. Give two advantages and two disadvantages of ICs.

4. A resistor is an active device. True or False?

5. What is another name for digital?

6. Explain the process of making integrated circuits.

7. Industry can purify silicon to_____ percent of purity in making ICs?

8. The cost of ICs has been going_____ (up or down) in the past few years?

9. Indicate in the basing diagram below which is pin #3 in the IC.

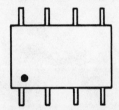

10. What is the digital number for binary number 0110?

11. In an OR GATE, if either SW_1 or SW_2 is closed, there will not be an output. True or False?

12. Name four input/output devices used in computers.

13. Explain the purpose of the arithmetic and logic unit (ALU) in a computer.

14. Some computer systems are designed for use in small businesses, _____ and _____.

15. In computer language, exterior units such as video displays are often called_____.

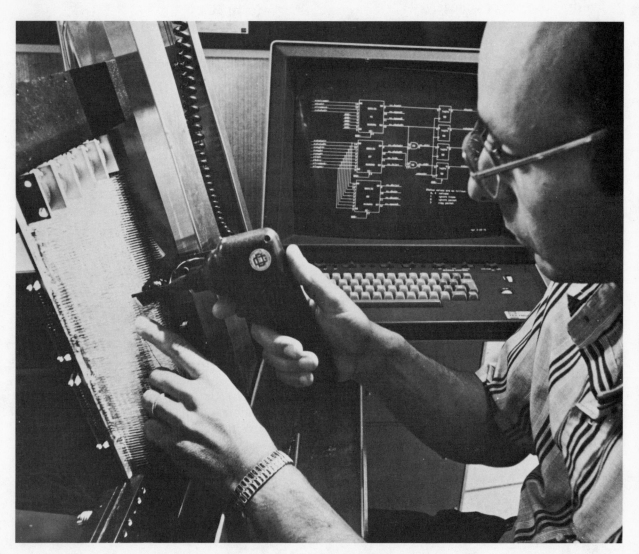

Computer systems research engineer uses a computer-controlled wire-wrap machine to build a prototype electrical circuit.   (Bell Laboratories)

Unit 12

ELECTRONICS IN INDUSTRY

This Unit explains how the field of electronics is used in two basic types of industries. Specific topics covered include:

1. How we communicate.
2. How a basic oscillator works.
3. What amplitude modulation (AM) is.
4. How a frequency modulation (FM) radio operates.
5. What the basic principles of television are.
6. How motors work.
7. What basic electronic controls are used in industry.

ELECTRONICS IN THE COMMUNICATIONS INDUSTRY

Communications is the process of giving and receiving information, signals and messages. Over the centuries, human beings have had to communicate. The earliest cave dwellers had to develop a language system in order to communicate. Although many devices and methods have been used to communicate by sound, the voice is the most universal communications medium used by people.

Pictographic writing, which uses pictures as communication symbols, was the early form of written messages. Then, the alphabet enabled people to communicate by written symbols. The printing press with movable type brought about an information explosion. The typewriter, or writing machine as it was called earlier, is also considered to be a communications tool.

Through the centuries, the distances over which we communicate have been increasing.

Early means were by gestures, drums, fire and smoke. The semaphore and flag system were used primarily on ships for communications purposes. The telegraph was one of the first devices to provide instant communications over great distances.

The telephone was the first long distance device invented to both transmit and receive the human voice. Radio, motion pictures and television are electronic communications mediums. Within the twentieth century, these devices and many others have become common methods of communication.

COMMUNICATION SYSTEM THEORY

Humans have been studying communications for centuries. In the past 50 years, however, communications system theory has been developed. One of the leaders in this field, Claude E. Shannon, published a lengthy paper on communications. Fig. 12-1 shows a block diagram of the communications system. An explanation of the elements of the diagram follows:

MESSAGE SOURCE: The message (information to be communicated) starts at the Message Source Block, Fig. 12-1. This message source could be an idea in a person's mind or in the author's mind. In the case of a computer sending or communicating information to another computer, the data is the message source.

CODER: The coder accepts the message from the source and changes the message into some

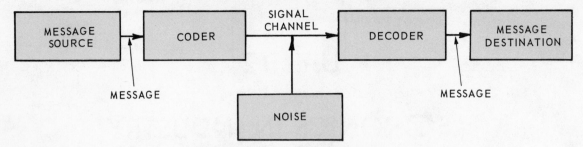

Fig. 12-1. Communications block diagram.

other form for tranmission. The transmitter at a radio station, a telegraph key and a microphone are all coders. For people, the voice is the primary coding device for transmitting messages to others. Also, we use our appearance, our touch and other body coders to transmit messages. Everyone has sensed love by a warm pat on the back, a handshake or a kiss.

CHANNEL: The channel is the link or medium over which the coded message is sent. The printed page is a coded message, and light rays entering your eyes make a signal channel. In radio, the transmitted electromagnetic signal is the channel. In voice communications, the coded words are transmitted through an air channel to the listener's ear.

DECODER: The decoder receives the transmitted message from the channel and reproduces it so that it can be understood at the message destination. In a radio, for example, the receiver is the decoder. The antenna picks up the signal, which is tuned, demodulated, amplified and reproduced as air waves by the speaker. Ears and eyes are decoders which convert sense inputs into impulses that are fed to the brain. Other sense inputs, such as touch, smell, taste, sight and hearing, likewise convert their signals into impulses for the brain.

MESSAGE DESTINATION: When the intended person, animal or machine has received the message, the message destination has been reached. The person, animal or thing receiving the message has to interpret it and give meaning to it. If the word "love" is transmitted, the person at the message destination must give individual meaning or understanding to the word. If the person is mean or ill, the transmitted love signal may be interpreted as HATE. The

background of the receiver determines how the received signal will be interpreted.

NOISE: Noise is anything which distorts the transmitted signal. A distorted signal may not properly deliver the message. An example of noise in a radio is static interferring with a signal. The effect of a weak television signal producing "snow" on the TV screen is a form of communications noise. Laughter, shouts or anything that takes the speaker's mind off the communications process can be considered noise. Improper light on this page can be a form of communications noise if the printed signal is not properly communicated to the reader's eyes and brain.

Fig. 12-2 shows two familiar communications processes. Each process is broken down into its various stages.

COMMUNICATIONS IN INDUSTRY

Norbert Wiener stated in his book THE HUMAN USE OF HUMAN BEINGS that "We ordinarily think of communication . . . as being directed from person to person. However, it is quite possible for a person to talk to a machine, a machine to a person, and a machine to a machine.

This profound statement explains how communication takes place in many industries. Fig. 12-3 shows two engineers "talking" to a line printer which is connected to a computer by telephone.

THE OSCILLATOR

The German scientist, Heinrich Hertz, discovered that an electric spark would set up a

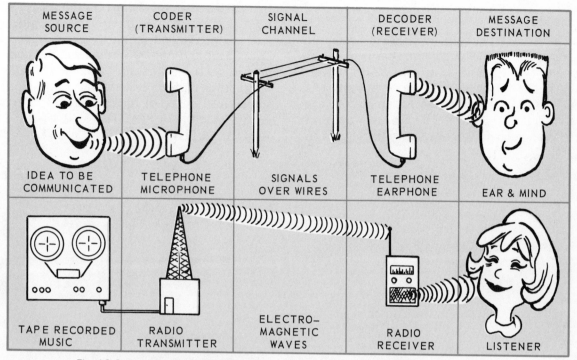

MESSAGE SOURCE	CODER (TRANSMITTER)	SIGNAL CHANNEL	DECODER (RECEIVER)	MESSAGE DESTINATION
IDEA TO BE COMMUNICATED	TELEPHONE MICROPHONE	SIGNALS OVER WIRES	TELEPHONE EARPHONE	EAR & MIND
TAPE RECORDED MUSIC	RADIO TRANSMITTER	ELECTRO-MAGNETIC WAVES	RADIO RECEIVER	LISTENER

Fig. 12-2. Two familiar communications methods and the stages required for process.

Fig. 12-3. Person-to-machine communications. (United Telecommunications Inc.)

vibrating, high frequency electric and magnetic field, and these vibrations could be picked up at some distance by a coil of wire. Thus was born the possibility of communications by vibrating electric waves.

Not until 1895 did the Italian scientist, Guglielmo Marconi, make the practical application of the Hertz waves. He communicated with the Italian Navy twelve miles out to sea by vibrating waves. The first message sent overseas to America startled the world in 1901.

We have already studied how alternating waves are generated by means of the alternator (Unit 8). However, the frequency of these waves is limited by the mechanics of the machine. You can make a rotating armature spin just so fast or it will fall apart. Higher frequency waves are required for radio communications.

Electronic circuits are used to generate these waves. They are called OSCILLATORS. To oscillate means to swing back and forth or to vibrate. A playground swing is a good example of a swinging or oscillating movement.

HIGH FREQUENCY WAVES

In electronics, we want a voltage or current to start at zero, rise to maximum in one direction or polarity, return to zero, then rise to maximum in the opposite direction or polarity and return to zero. We want this action to continue at some high frequency. The desired wave produced is similar to the one shown in Fig. 12-4.

the current also will change continuously. So we must have a way of comparing an ac value with a dc value. Since current through resistance produces heat and does work, we will discover the answer to the question: "What peak value of alternating current will do the same work as some fixed value of direct current?" This has been named the EFFECTIVE VALUE OF THE ALTERNATING CURRENT WAVE.

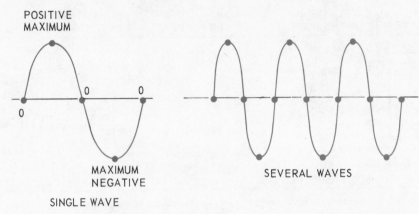

Fig. 12-4. Wave shows the varying voltage or current through cycle or alternation.

DESCRIPTION OF THE WAVE

So that we can talk about this high frequency wave and understand each other, we will put some names on the wave values and actions. See Fig. 12-5. The PEAK value of the wave is from zero to the peak value in one direction. The PEAK-TO-PEAK value is from maximum in one direction to maximum in the opposite direction.

The FREQUENCY of the wave is the number of times the wave goes through its cycle per second. It is measured in Hertz. The PERIOD of the wave is the time in seconds it requires for one cycle. For example, a 100 Hertz wave requires 1/100 second for one complete cycle or oscillation.

EFFECTIVE VALUE

We learned early in our studies that current in a circuit is caused by voltage or pressure. In the alternating voltage wave, the voltage never remains at a set value. It is changing continuously. If such a generator is connected to a circuit,

Actually, the effective value of an alternating current or voltage is what you usually measure with a meter. The effective value also may be found by simple arithmetic:

$$\text{PEAK VALUE} \times .707 = \text{EFFECTIVE VALUE}$$

or

$$\text{EFFECTIVE VALUE} \times 1.414 = \text{PEAK VALUE}$$

AN OSCILLATING CIRCUIT

Examine the circuit in Fig. 12-6. C is originally charged when SW_1 is closed. When SW_1 is opened and SW_2 is closed, the capacitor discharges through the coil and produces a magnetic field. After C has discharged, the collapsing field charges C in the opposite direction. The current oscillates back and forth in the circuit until the ENERGY originally stored in C from the power source is used up.

Resistance in the circuit uses up a little energy with every current swing. Compare this to a

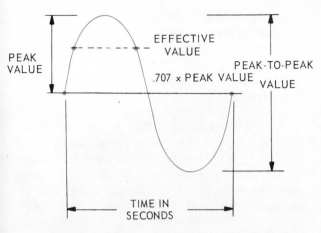

Fig. 12-5. Parts of a wave. The effective value of the wave compares the alternating current wave to a direct current value.

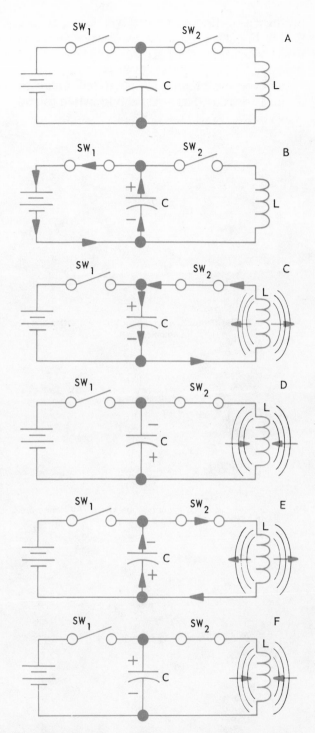

Fig. 12-6. Follow the action of an energized LC circuit. A—SW₁ open, SW₂ open. B—SW₁ closed, SW₂ open, C charges. C—SW₁ open, SW₂ closed, C discharges through L and produces magnetic field. D—When C is discharged, the magnetic field collapses and charges C in opposite polarity. E—C now discharges in opposite direction and produces magnetic field. F—When C is discharged, the magnetic field collapses and charges C back to its original polarity. ONE OSCILLATION IS COMPLETE.

playground swing. Each swing becomes less and less until the swing comes to rest. A current wave, gradually decreasing in height, is shown in Fig. 12-7. It is a DAMPED WAVE that becomes less and less with each oscillation.

TUNED TANK CIRCUIT

The previous circuit of the capacitor and a coil is called a TANK circuit and the oscillating current is compared with water sloshing back and forth in a tank. The current action is sometimes referred to as FLYWHEEL ACTION. The important point you must realize is: SOME VERY DEFINITE PERIOD OF TIME MUST PASS DURING WHICH THE CIRCUIT COMPLETES ONE OSCILLATION. This time depends upon the VALUE OF L and C. For any set pair of values, the circuit will oscillate at a particular frequency. To change the frequency of oscillation, it is only necessary to change either the value of C or L. This is called TUNING THE CIRCUIT.

CONTINUOUS WAVES

On the playground, the swing will continue to swing back and forth, if only a little push is added when the swing is at its backward position. The push replaces the energy used up by the friction of the swing. The same is true of the oscillating tank circuit. If a little pulse of electrical energy is added at the correct frequency, then the oscillator will continue to oscillate and produce a continuous wave (CW). Compare the CW in Fig. 12-8 to the damped wave shown in Fig. 12-7.

In electronic circuits, an integrated circuit or a transistor is used as a fast acting switch to add

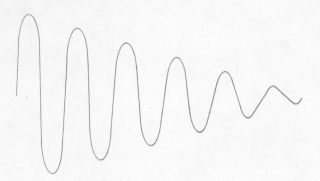

Fig. 12-7. The damped wave becomes less with each oscillation.

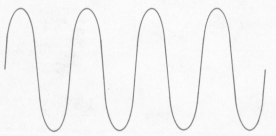

Fig. 12-8. Four cycles of a continuous wave. All peaks are equal.

energy to the tank circuit at exactly the right moment. This is the oscillator circuit. An amplifier feeds pulses of energy to the tank circuit. A small part of the energy in the tank circuit is fed back and reamplified to keep the tank circuit oscillating.

EXPERIENCE 1. Build an AUDIO OSCILLATOR. This circuit produces a tone you can hear. Many young students build one for practicing the Morse code such as used in ham radio work or Scouting. The circuit diagram appears in Fig. 12-9 and the pictorial in Fig. 12-10. A key may be connected in the emitter lead to key the oscillator to sounds required for code practice. Turn the potentiometer control and notice how it changes the pitch or frequency of the oscillator.

EXPERIENCE 2. Connect your oscilloscope across the speaker terminals. Adjust the oscilloscope for three stationary waves about one inch high. Change the frequency of the oscillator and observe that the oscilloscope frequency must also be changed.

EXPERIENCE 3. Replace the 10 μF capacitor with a .1 μF paper capacitor. Now the pitch is very high and sharp. Many electronic organs use circuits similar to these for producing the notes in the musical scale. This type of circuit is also used to test your hearing. Usually,

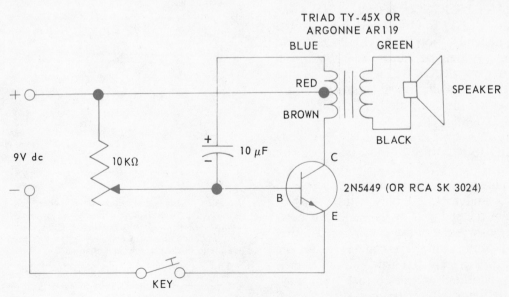

Fig. 12-9. The schematic diagram for the audio oscillator.

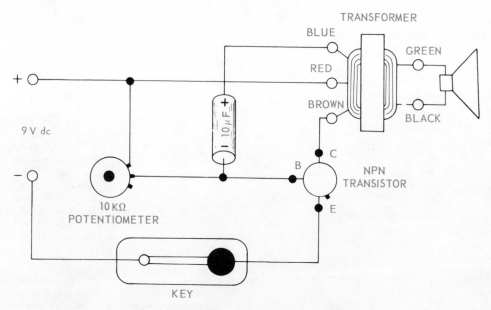

Fig. 12-10. Pictorial diagram of audio oscillator.

humans can hear up to 18 KHz. As you grow older, your hearing ability decreases. Dogs and animals sometimes hear sounds above 20 KHz.

RF OSCILLATORS

An oscillator which operates at frequencies above 300,000 Hz is generally classed as a radio frequency oscillator. Oscillators of this frequency and higher are attached to the radio transmitting antenna and the generated waves radiate into space. These waves (received from the antenna on the roof of your house) bring you the sounds of music and voice and television pictures.

RF AMPLIFIERS

The simple oscillator usually does not have enough power to generate a strong wave, and it would not radiate any great distance. To overcome this and assure you will receive the radio program, the oscillator output will be amplified to high power by transistors and tubes before it is coupled to the transmitting antenna.

CODE COMMUNICATIONS

A very reliable way of transmitting radio messages is by using the Morse code. Of course, you must be able to understand the message.

The code is used extensively by amateur and commercial radio stations.

For code communications, the radio transmitter is turned on or off by a hand switch or key. A short burst of radiated energy represents a "dot" or "di" or "dit." A longer burst a "dash" or a "dah." For example, a short and a long burst, "dit dah" (written · -) means the letter A. Fig. 12-11 shows the distress call ("SOS") for ships at sea, written in code with the radio frequency waves above. The complete MORSE CODE used for this purpose may be found in Fig. 12-12.

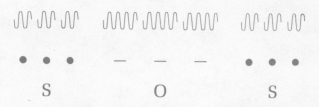

Fig. 12-11. SOS written in code with radio waves. A short burst is a "dit"; a longer burst a "dah."

IMPORTANT: THE MORSE CODE IS SENT OVER THE AIR AT A FIXED FREQUENCY. TO RECEIVE THE CODE, THE RADIO RECEIVER MUST BE TUNED TO THE SAME FREQUENCY.

A	• —	(di–dah)
B	— • • •	(dah–di–di–dit)
C	— • — •	(dah–di–dah–dit)
D	— • •	(dah–di–dit)
E	•	(dit)
F	• • — •	(di–di–dah–dit)
G	— — •	(dah–dah–dit)
H	• • • •	(di–di–di–dit)
I	• •	(di–dit)
J	• — — —	(di–dah–dah–dah)
K	— • —	(dah–di–dah)
L	• — • •	(di–dah–di–dit)
M	— —	(dah–dah)
N	— •	(dah–dit)
O	— — —	(dah–dah–dah)
P	• — — •	(di–dah–dah–dit)
Q	— — • —	(dah–dah–di–dah)
R	• — •	(di–dah–dit)
S	• • •	(di–di–dit)
T	—	(dah)

U	• • —	(di–di–dah)
V	• • • —	(di–di–di–dah)
W	• — —	(di–dah–dah)
X	— • • —	(dah–di–di–dah)
Y	— • — —	(dah–di–dah–dah)
Z	— — • •	(dah–dah–di–dit)

1	• — — — —	(di–dah–dah–dah–dah)
2	• • — — —	(di–di–dah–dah–dah)
3	• • • — —	(di–di–di–dah–dah)
4	• • • • —	(di–di–di–di–dah)
5	• • • • •	(di–di–di–di–dit)
6	— • • • •	(dah–di–di–di–dit)
7	— — • • •	(dah–dah–di–di–dit)
8	— — — • •	(dah–dah–dah–di–dit)
9	— — — — •	(dah–dah–dah–dah–dit)
0	— — — — —	(dah–dah–dah–dah–dah)

Period	• — • — • —	(di–dah–di–dah–di–dah)
Comma	— — • • — —	(dah–dah–di–di–dah–dah)
Question	• • — — • •	(di–di–dah–dah–di–dit)

Fig. 12-12. The MORSE CODE.

THE COMMUNICATIONS SYSTEM

A simplified BLOCK DIAGRAM of the transmitter and receiver is shown in Fig. 12-13. This method of representing circuits is widely used in industry, so that the technician can observe the total system and the individual units in the system.

TUNING AN AM STATION

You may wish to review our discussion of tuned circuits covered earlier in this Unit. Then, see Fig. 12-14 for a diagram of a typical radio tuning circuit. In this diagram, the capacitor is the variable type, controlled by a knob on the front panel of your radio. By varying the capacitance, the circuit will oscillate at any frequency between 550 KHz and 1600 KHz.

If a radio station is broadcasting at 980 KHz, it is only necessary to tune the receiver to 980 KHz to hear the station. All radios, including FM and television, operate on this same principle. However, they use different frequencies and, therefore, different values of L and C in their tuned circuits.

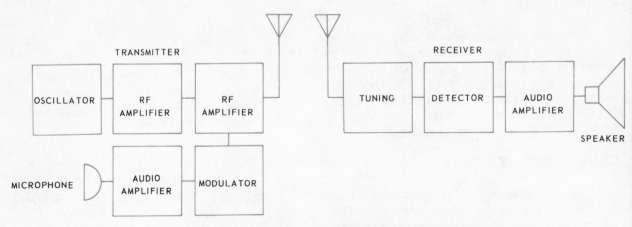

Fig. 12-13. The block diagram of a simplified radio transmitter and receiver.

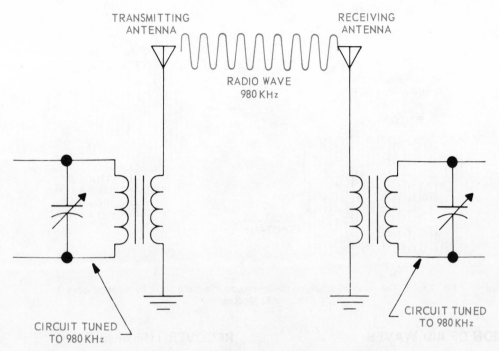

TRANSMITTING ANTENNA

RECEIVING ANTENNA

RADIO WAVE 980 KHz

CIRCUIT TUNED TO 980 KHz

CIRCUIT TUNED TO 980 KHz

Fig. 12-14. The radio receiver is tuned to the same frequency as the transmitted signal.

Consider, again, the swinging playground swing. By pushing it at its most backward swing at the right time, you caused the swing to continue. Suppose you pushed the swing forward as it was moving backwards. Then, you would oppose the swinging motion and the swing would quickly stop. The swing must be pushed at the same frequency or multiples of that frequency to keep the swing going.

Likewise, the receiver tuned circuit must be adjusted by varying the capacitance or inductance so that the circuit will swing or oscillate when energy is added from the transmitted radio wave.

AMPLITUDE MODULATION

You would not enjoy a radio program sent by the Morse code. In fact, you would need a lot of practice to understand it. In order to send music and voice over the radio, something special must be done. This is accomplished by MODULATION. Instead of transmitting a constant amplitude radio frequency wave, we must add circuits to the transmitter which will make the radio wave amplitude vary to match the frequency of voice or music. This is classed as AMPLITUDE MODULATION (AM).

Your broadcast band radio receives AM waves. To better picture such a wave, study Fig. 12-15. Here, a 400 Hz sound wave causes the radio frequency wave to change its height or amplitude at the rate of 400 Hz. This modulated radio wave now carries a 400 cycle tone as it is broadcasted into space.

Many audio tones and speech can be used which result in a very complex modulated wave.

EXPERIENCE 1. Connect your oscilloscope to the output of a radio frequency generator. Set generator frequency at about 1000 KHz to produce a green band across the oscilloscope face about one inch in amplitude. Set the oscilloscope frequency near 400 Hz. Turn on the modulation in the RF generator and observe the modulated pattern. Slight adjustment of oscilloscope frequency may be necessary. Does it appear similar to the modulated radio wave at right in Fig. 12-15?

EXPERIENCE 2. Without changing the oscilloscope controls, connect it across the speaker of any AM radio while a radio program is being received. Now, you will see the complex pattern resulting from tones and sounds of many frequencies.

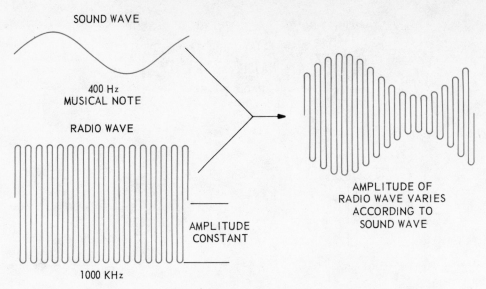

SOUND WAVE

400 Hz
MUSICAL NOTE

RADIO WAVE

AMPLITUDE
CONSTANT

1000 KHz

AMPLITUDE OF
RADIO WAVE VARIES
ACCORDING TO
SOUND WAVE

Fig. 12-15. The result of changing the amplitude of a radio wave by a sound wave is Amplitude Modulation.

DETECTION OF AM WAVES

If you tuned in an AM wave that was connected directly to a speaker or earphones, you would hear nothing. Looking at the modulated wave again, you will see that for every positive peak there is an equal negative peak, and the average value of the wave is zero. In order to detect the variation in amplitude of the wave (music or speech), one-half of the wave must be removed by rectification. In this application, it is called DETECTION. A diode is used in Fig. 12-16, and both input modulated wave and the demodulated or detected output are illustrated. Only the positive half of the modulated wave causes an output since the diode cuts off the negative half of the cycle. Note: You could get a negative output by reversing the diode. This is sometimes done.

RECOVER THE MUSIC

In Fig. 12-17, the detected AM radio wave is again shown. It consists of pulses of positive voltage of various amplitudes. ANY cycle is at its peak for only a fraction of a second. At all other times, the detected AM radio wave shown on the oscilloscope continuously reads at some value between peak and zero. Therefore, the wave must have an AVERAGE VALUE, which is indicated in Fig. 12-17.

Observe that this VARYING AVERAGE VALUE wave has the same appearance as the original sound wave used to modulate the transmitter shown in Fig. 12-15. Now, it can be amplified and used to drive a speaker. The radio frequency wave remaining can be filtered out and bypassed to ground.

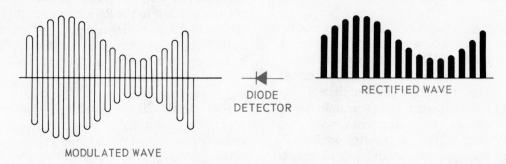

DIODE
DETECTOR

RECTIFIED WAVE

MODULATED WAVE

Fig. 12-16. After rectification, only the positive cycles of the wave remain. When modulated, wave is negative. The diode does not conduct.

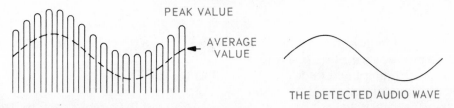

PEAK VALUE

AVERAGE VALUE

THE DETECTED AUDIO WAVE

Fig. 12-17. The average value of the detected RF wave is the original sound wave.

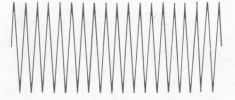

UNMODULATED 100 MHz RF WAVE

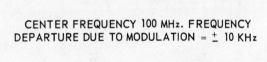

AUDIO WAVE

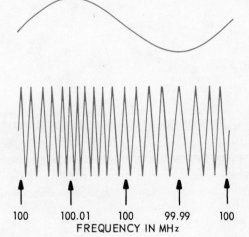

CENTER FREQUENCY 100 MHz. FREQUENCY
DEPARTURE DUE TO MODULATION = ± 10 KHz

100   100.01   100   99.99   100
FREQUENCY IN MHz

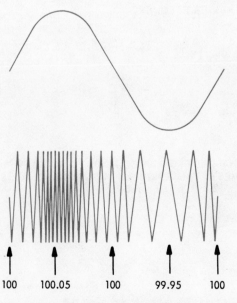

AUDIO WAVE
GREATER AMPLITUDE

CENTER FREQUENCY 100 MHz. FREQUENCY
DEPARTURE DUE TO MODULATION = ± 50 KHz

100   100.05   100   99.95   100

Fig. 12-18. Strength of amplitude of sound wave determines amount of departure from center frequency.

FM RADIO

AM radio has its limitations. When high fidelity music is broadcasted, the AM radio wave is also sensitive to noise and interference. In the frequency modulated or FM radio system, the radio frequency wave is kept at a constant amplitude. The modulating sound wave causes the radio wave to change frequency at the same rate as the audio wave. The strength or loudness of the audio wave determines the amount of frequency change. Each of these conditions is described in Figs. 12-18 and 12-19.

The FM broadcast band extends from 88 MHz to 108 MHz. In the example, the CENTER FREQUENCY of 100 MHz is used.

FM DETECTION

Although the circuits used to detect frequency variation are covered in more advanced textbooks, you should realize that this method of sending information over radio waves is distinctly different. A detector for AM must be sensitive to amplitude changes in a signal; a detector for FM must be sensitive to frequency

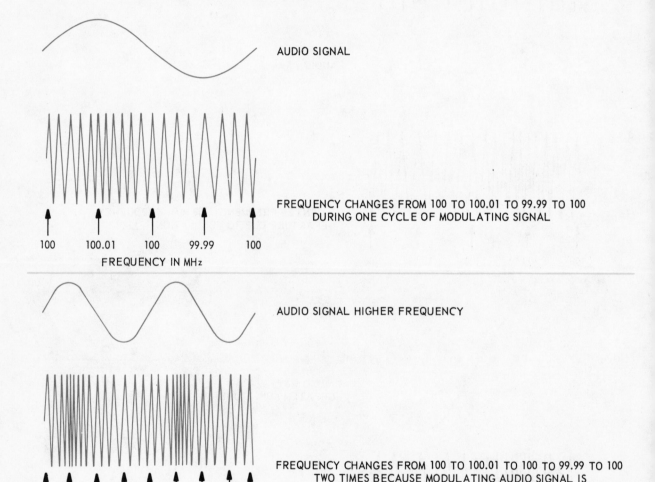

AUDIO SIGNAL

FREQUENCY CHANGES FROM 100 TO 100.01 TO 99.99 TO 100
DURING ONE CYCLE OF MODULATING SIGNAL

100   100.01   100   99.99   100
FREQUENCY IN MHz

AUDIO SIGNAL HIGHER FREQUENCY

FREQUENCY CHANGES FROM 100 TO 100.01 TO 100 TO 99.99 TO 100
TWO TIMES BECAUSE MODULATING AUDIO SIGNAL IS
2 TIMES THE ORIGINAL SIGNAL

100   100   100   100   100
   100.01   100.01
        99.99   99.99
FREQUENCY IN MHz

Fig. 12-19. The RATE OF FREQUENCY change depends upon the frequency of the audio modulating signal.

changes. Each method must recover the sound waves used originally at the transmitter for modulation.

FM MULTIPLEXING

A tremendous advantage of FM radio is the capability of sending more than one radio program over the same frequency carrier wave. The familiar example is STEREOPHONIC FM where the LEFT channel is separated from the RIGHT channel. At the receiver they are detected and amplified separately and connected to LEFT and RIGHT speakers. This produces a musical front which gives the same effect as listening to a symphonic orchestra in a theater, if this kind of music is being broadcasted.

TELEVISION

What is a picture? Examine a photograph in a newspaper or magazine with a magnifying glass. You will see that the picture is made up of thousands of small dots or areas varying from white to black (or color) and all shades in between. These are called picture elements. The arrangement of these elements makes the picture.

The video camera at the studio looks at the scene through a lens similar to a camera. Instead of film, the video camera uses a photo sensitive plate of many photo cells. Each element focused onto the plate produces a voltage corresponding to the darkness or lightness of the picture element. An electron beam in the video camera scans (looks over in detail from point to point) the photo sensitive plate. The electron beam moves from left to right and downward in the same fashion as you are reading this printed page. It reads very fast however. In fact, the picture contains 525 lines and the beam reads these lines 30 times per second. It therefore produces 30 complete pictures per second. The lightness or darkness of the picture element is changed into an electrical signal. The darker elements produce larger signals.

Fig. 12-20 shows a checker board and the signal developed in the camera as the electron beam scans two lines of the picture. The signal is used to modulate an AM transmitter in the same manner as for radio broadcasting.

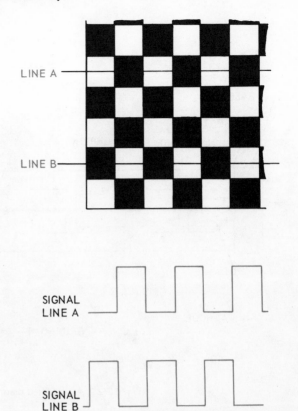

Fig. 12-20. The camera signal developed as electron beam scans two lines of a checkerboard.

PICTURE TUBES

A simplified drawing of a television picture tube is shown in Fig. 12-21. In operation, it is similar to a vacuum tube. The HEATER and the CATHODE produces the space cloud of electrons. GRID No. 1 is the control grid and the number of electrons passing through it depends upon its negative voltage. GRIDS 2, 3, 4, and 5 are used for acceleration of the electrons toward the screen and for focusing the electron beam to make a tiny spot.

When the electron beam hits the fluorescent screen, it causes the screen to glow at the degree of whiteness depending on the intensity of the beam. This is controlled by GRID No. 1. The electrons are collected by the high voltage anode to complete the circuit.

SCANNING

Around the neck of the picture tube are magnetic coils, Fig. 12-21, called the DEFLECTION YOKE. By varying the current through

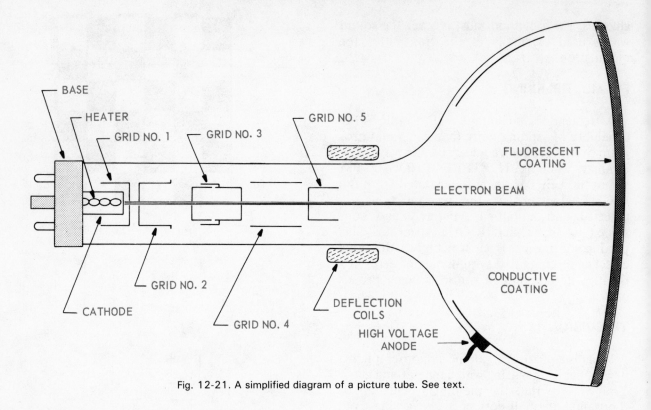

Fig. 12-21. A simplified diagram of a picture tube. See text.

these coils, the beam can be made to move back and forth (horizontal scanning) and upward and downward (vertical scanning). In the actual tube, the beam starts at the top left-hand corner, Fig. 12-22, and moves across the screen, left to right. At the extreme right position, the beam is turned off and caused to RETRACE to the left position again. It is now moved

downward slightly and made to trace another line in the same manner. This beam traces out 15,750 horizontal lines per second. In American television 525 lines are used to form one complete picture and the picture is scanned 60 times per second.

As the beam moves from top to bottom, it first traces the lines 1, 3, 5, etc., or odd lines. When it reaches the bottom, it rapidly returns to the top again and traces out the even numbered lines. This is referred to as IN-TERLACE SCANNING. It produces a better picture. One scan of the even or odd lines is a FIELD. Complete coverage of the screen by both even and odd lines is a FRAME. There will be 60 fields per second or 30 frames per second.

SYNCHRONIZATION OR SYNC

You will recall that the television camera at the studio also had an electron beam scanning a mosaic photo sensitive plate and converting the darkness of the elements into electrical signals. If the home TV is to reproduce this same picture, the scanning rate in the home TV must match the scanning rate of the TV camera. They must be in step like a squad of marching

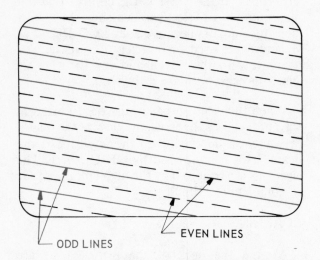

Fig. 12-22. The odd lines are traced first and then the even lines. There are 262 1/2 odd lines and 262 1/2 even lines in the total picture.

soldiers. To accomplish this feat, the TV transmitter sends a SYNCHRONIZATION (SYNC) PULSE to the home TV set for each line traced on the screen. This pulse keeps the horizontal and vertical sweep oscillator circuits exactly on frequency.

The signal required for two lines of a TV picture as transmitted over the air is shown in Fig. 12-23. It is the composite VIDEO SIGNAL. Note the varying voltages (for light or dark) applied to grid No. 1 of the picture tubes as it traces one line on the screen. At the end of the line appears a BLANKING PULSE, which turns off the beam during its retrace or return time. On top of the blanking pulse is the SYNC PULSE, used to keep the home TV in step with the TV station.

Also note that the greater the amplitude of the video information signal, the blacker the spot on the screen appears. During the blanking pulse, the screen is completely black. However, it happens too fast for the human eye to detect.

TV SOUND

The television station uses a typical FM transmitter for broadcasting the sound. This is an independent transmitter, and it should always be in step with the picture. Once in a while something may go wrong, like a dog barking with a soprano voice. Mistakes do happen.

EXPERIENCES WITH YOUR TV

There are two types of adjustments which are made on TV sets. The first group involves technical corrections that should be made by your TV service technician. The second group consists of OWNER CONTROLS. These are the knobs on the front and/or side panel of the TV.

CHANNEL SELECTION. This control is a ROTARY switch. Each channel is transmitted at a fixed frequency. The channel selector switches in the correct tank circuit (review tuning material on page 157) to respond to the channel frequency.

FINE TUNING. The fine tuning knob makes minor adjustments on the tuning circuit, so that your TV is exactly tuned to incoming signals.

BRIGHTNESS CONTROL. This knob changes the grid voltage of the picture tube and makes a brighter or darker overall picture.

CONTRAST CONTROL. This control changes the gain of the amplifiers to make a greater difference between black and white.

HORIZONTAL HOLD. This adjustment makes slight changes in the horizontal scanning oscillator frequency so that it will lock in step with the sync pulses.

VERTICAL HOLD. This control adjusts the vertical scanning oscillator so that it will lock in step with the vertical sync signal.

FOCUS CONTROL. This knob controls voltages on picture tube to cause electron beam to converge to a narrow beam.

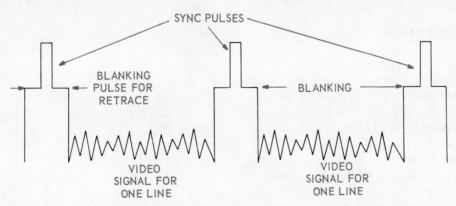

Fig. 12-23. The composite video signal showing video information, blanking pulse for retrace and sync pulse.

VOLUME CONTROL. This knob controls the level of sound. Usually, the ON-OFF SWITCH is part of this control.

COLOR TELEVISION

Color television produces a more realistic picture than monochrome (black and white) television. A color picture consists of the PRIMARY COLORS OF RED, GREEN AND BLUE in various combinations. All natural colors can be made by mixing these primary colors. In the television studio, the camera separates the scene according to its color content. Optical filters send the red, green and blue colors to separate camera tubes where they are scanned and converted into signals similar to the black and white camera.

These individual signals are not broadcasted as they are to your home, for a good reason. According to the FCC regulations, a black and white TV must be able to receive a color program and produce it as black and white. Therefore, the individual color signals are combined to form a black and white signal and a composite color signal. These two signals are sent to the home receiver where they are separated to make the color picture.

The black and white signal is made by combining the colors to indicate the variations in brightness of the picture. It is called the LUMINANCE SIGNAL. The colors in required proportions are mixed to form the CHROMINANCE SIGNAL. Both signals are used to modulate the transmitter. Multiplexing methods are used.

COLOR TV RECEPTION

Your home TV receiver demodulates or detects the signals and decodes them. The three colors in proper proportions are fed to the picture tube. The color TV tube has three electron guns. One for each color. The screen of the color tube is phosphor coated with trios of red, blue and green phosphors. Each of the three color electron guns is directed to make its respective dot in the color trio glow. The total picture is a combination of all colors in varying degrees.

When speaking of a color TV picture, the term HUE is used. This is the color of the picture as the eye sees it. Grass has a green hue. Our flag has hues of red, blue and white. WHITE is a proper mixture of red, green and blue. SATURATION is an expression of how rich the color is. A color mixed with white can be washed out. On the other hand, the red in our flag is very rich and vivid. Your color TV will have adjustments for hue and saturation.

ELECTRONICS IN THE MANUFACTURING INDUSTRIES

The conversion of ELECTRICAL ENERGY TO MECHANICAL ENERGY is made possible by the ELECTRIC MOTOR. To understand the principle of the motor, we need a basic understanding of the reaction between two magnetic fields.

MAGNETISM AND MOTORS

In Unit 7, the following laws of magnetism were explained:

UNLIKE MAGNETIC POLES ATTRACT
LIKE MAGNETIC POLES REPEL

Later, we discovered that a magnetic field existed around a current-carrying conductor, and the direction of this field depended on the direction of current flow.

In Fig. 12-24, a current-carrying conductor is shown in a magnetic field. Current is flowing

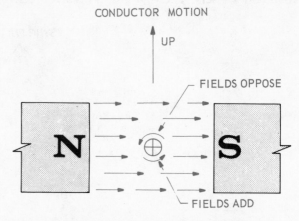

Fig. 12-24. Above the conductor, the fields oppose and are weakened. The conductor moves upward.

"in" as indicated by the cross on the end of the wire. The circular magnetic field around the conductor illustrates this action. Above the conductor, the fixed field and the conductor field oppose each other. The conductor moves upward or toward the weakened field.

In Fig. 12-25, the opposite action takes place as current is flowing outward in the conductor as indicated by the dot. Now the field is weakened below the conductor, and the conductor moves downward. This is called MOTOR ACTION. It is this principle which makes a motor rotate.

DC MOTORS

A simple motor is sketched in Fig. 12-26. It has a permanent magnetic field. The conductor is in the shape of a coil called the ARMATURE. The armature is mounted on a shaft so it can rotate on bearings.

Note that the ends of the armature coil are connected to sections of a metal ring. This is the

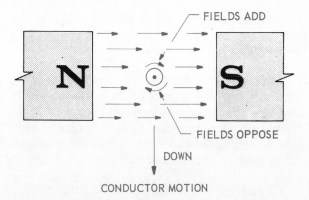

Fig. 12-25. The fields oppose below the conductor and are weakened. The conductor moves downward.

COMMUTATOR. It is mounted on the same shaft and revolves with the armature. Outside power is brought into the armature through brushes rubbing on the commutator sections.

In the first position, follow the current flow from outside power through the brushes and commutator to the armature and out the other side. Current flows INWARD in armature coil marked C. Motor action causes an UPWARD

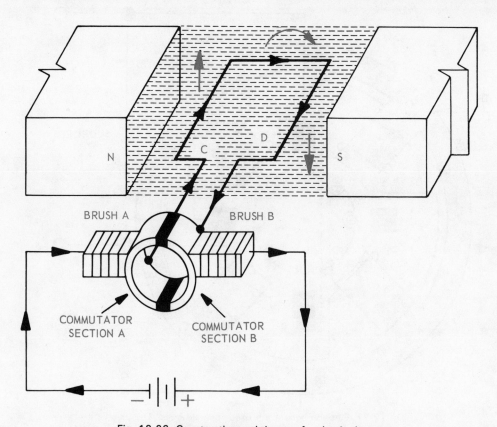

Fig. 12-26. Construction and theory of a simple dc motor.

movement. Current flows OUTWARD in armature coil marked D. Motor action causes a DOWNWARD movement, and the coil turns.

After one-half revolution, the commutator section B comes in contact with brush A, which reverses the armature current. Then, side D moves upward and side C moves downward. The switching action of the commutator causes the armature to continue rotation.

To increase the turning power of the motor, the armature may have many coils connected to commutator sections, and the field magnets may be electromagnets. A more practical motor is illustrated in Fig. 12-27.

THE MOTOR AS A GENERATOR

When you compare the simple dc motor to the dc generator in Unit 8, you will discover that they are quite similar. In fact, power applied to a generator will make it a motor. On the other hand, a motor is a generator. As the armature windings rotate through the magnetic field, they generate a voltage which opposes the voltage from the source. This generator voltage is COUNTER ELECTROMOTIVE FORCE (CEMF).

We know that current in any circuit depends on the applied voltage. When a motor is not rotating, there is no CEMF and the current is the result of armature resistance only and source voltage.

$$I_{armature} = \frac{E_{source}}{R_{armature}}$$

When the motor starts to rotate, the generated CEMF opposes the source voltage. The net voltage becomes much less and the current decreases.

$$I_{armature} = \frac{E_{source} - E_{cemf}}{R_{armature}}$$

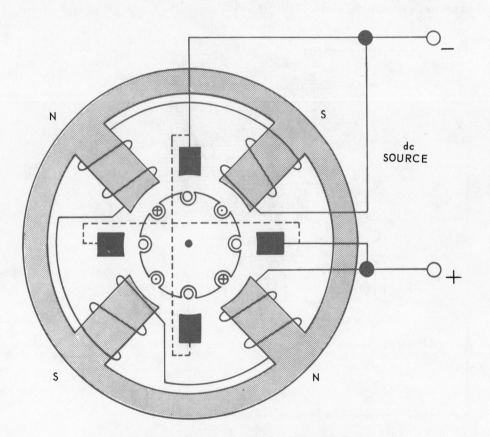

Fig. 12-27. Electromagnets are used for field coils. The armature has several coils connected to commutator sections.

CONCLUSIONS: A motor draws a lot of current when starting. The current becomes less as the motor builds up speed. At home, you may have noticed that your lights dim slightly when your refrigerator or other major appliance is turned on. Can you explain why?

AC MOTORS

The majority of motors use alternating current. This power is available from your power company. Will the dc motor operate on ac? With certain design changes, it can become a pretty good motor. If the field and the armature change magnetic polarity at the same time, the motor action is the same. Motors used for both dc and ac are called UNIVERSAL MOTORS. These are used on small tools such as drill motors, grinders and ventilating fans.

INDUCTION MOTORS

In the typical ac induction motor, the outside power is applied to the field or stator windings. By using inductance and capacitance with these windings, a rotating magnetic field can be created. The armature or ROTOR is a laminated, drum-like core with heavy single loop copper wires set in slots around its surface. Study Fig. 12-28 and notice this construction.

Fig. 12-28. A commercial ac induction motor. Note stator coils and construction of ROTOR. (Delco Products, GMC)

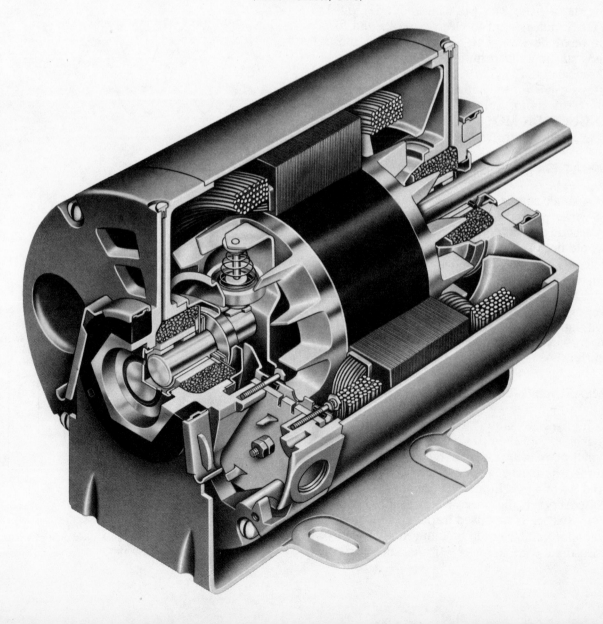

An induction motor can be considered a transformer. The stator windings will be the primary, and the shorted loops of heavy copper in the rotor are the secondary coils. The rotating magnetic field of the stator will induce a voltage in the rotor coil. The very low resistance of the single loop rotor coil will permit high currents to flow around the loops. These currents produce a magnetic field in the rotor with a polarity which will cause the rotor to follow around and turn with the rotating stator field. Tremendous turning power or TORQUE can be developed in this manner.

TYPES OF AC MOTORS

AC induction motors are usually named according to the method used to create the revolving stator field. You will hear about SHADED POLE motors used in electric clocks and low-powered devices. The SPLIT PHASE motor uses starting and running windings of unequal inductances. The CAPACITOR START motor uses a capacitor to change the phase of the ac to start the motor rotating. The POLYPHASE INDUCTION MOTOR uses ac from a 3 phase source to cause rotation.

WHAT IS PHASE?

A single source of ac power may alternate at 60 cycles per second as illustrated in Fig. 12-29. A second current could also be generated which would not rise and fall in step with the first current. This could be 2 phase. A more popular current used for large motors in industry is the 3 phase. This current is generated by coils in the generator 120 degrees apart. Each set of coils will have its own external connection. Three or four wires are required to conduct 3 phase current.

MEASURING RPM

Fig. 12-30 shows a rotating wheel of a machine. Attached to the rim of the wheel is a small permanent magnet. Each time the wheel turns, the magnet passes a small PICKUP coil and a voltage is induced in the coil (Faraday's Discovery). If the pulses occur in rapid succession (wheel is revolved rapidly), an average value of voltage could be induced. This can be measured with a meter. If the speed of the

rotating wheel were increased, the average value of the induced voltage would increase.

The meter, rather than reading voltage, could be marked in rpm (revolutions per minute). A constant check on the speed of the machine could be made. Such a circuit can be used for measuring the speed of many two cycle engines, which already have a magnet on the flywheel.

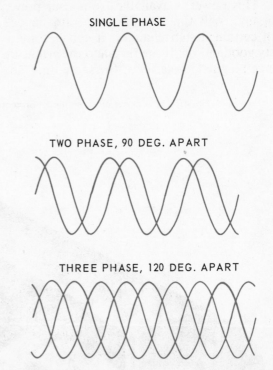

SINGLE PHASE

TWO PHASE, 90 DEG. APART

THREE PHASE, 120 DEG. APART

Fig. 12-29. The wave forms for single, two and three-phase generators.

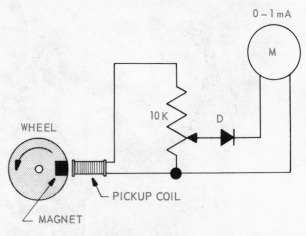

Fig. 12-30. A pulse is created for each revolution of wheel. The meter is calibrated in rpm.

A circuit similar to the rpm counter is used in the newer automotive electronic ignition systems. The coil and magnet are used instead of breaker points in the distributor. The pulse produced triggers a transistor or solid state switch and the current in the ignition coil. A high quality, intensive spark appears at the spark plug.

EXPERIENCE 1. Set up circuit pictured in Fig. 12-31. A magnet is used for a core of the coil. Move magnet in and out. Note meter reading. Remove capacitor. What purpose did it serve. Reverse diode. Why is diode used?

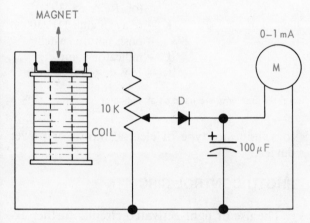

Fig. 12-31. The pictorial sketch for EXPERIENCE 1.

CONCLUSIONS: The increased movement produced a higher voltage. The meter could be read in "times per minute." The diode allows the current to flow in one direction only. The meter reads only in one direction. The capacitor seemed to average the voltage reading so that quick movements of the magnet did not cause the meter to move rapidly.

EXPERIENCE 2. Switching a transistor by a pulse. Build the circuit shown in Fig. 12-32. Move the magnet quickly in the coil. The lamp turns on. Does the transistor turn on when you move the magnet IN or OUT? Is a diode necessary in this circuit?

CONCLUSIONS: In order for the NPN transistor to conduct, the emitter must be negative and the base positive. This is forward bias for the NPN transistor. A voltage pulse of correct polarity will turn the transistor ON. A voltage

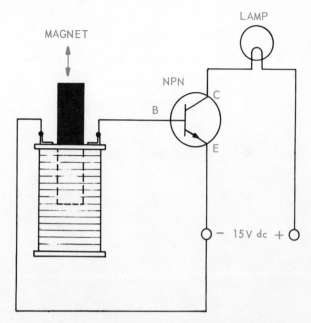

Fig. 12-32. The sketch for EXPERIENCE 2. The pulse from the magnet and the coil will turn on transistor. The lamp will light.

pulse of opposite polarity would only turn it more OFF. A diode is not necessary because the emitter-base junction serves that function in this circuit.

TIMING CIRCUITS

There are many occasions in industry when timing an operation or process is required. Timing circuits are used extensively. Even at home you may use a timer when taking pictures with your camera. In the darkroom, photographs are enlarged. The interval of exposure time may be controlled by the electronic timer. Your mother may use a timer when boiling eggs for breakfast.

Most electronic timing circuits use the time constant of a RESISTANCE-CAPACITANCE circuit to set the time interval. The method is quite accurate for short periods of time when quality components are used.

EXPERIENCE 3. Build a VARIABLE TIMING CIRCUIT. The circuit is shown in Fig. 12-33. Pressing SW_1 activates the circuit which energizes the relay and the indicator lamp turns ON. After an interval of time passes, adjusted by POTENTIOMETER R_2, the indicator lamp turns off.

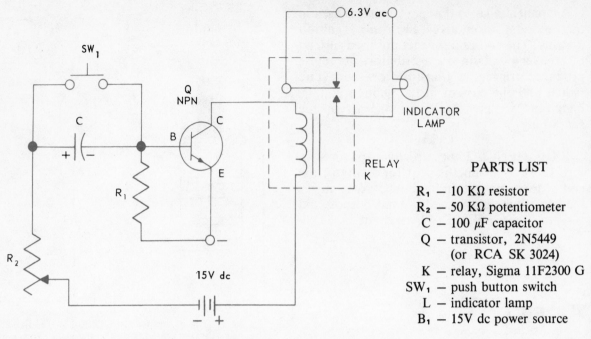

Fig. 12-33. The experimental TIMER circuit. The relay and lamp are used as an indicator.

PARTS LIST

R_1 — 10 KΩ resistor
R_2 — 50 KΩ potentiometer
C — 100 μF capacitor
Q — transistor, 2N5449
    (or RCA SK 3024)
K — relay, Sigma 11F2300 G
SW_1 — push button switch
L — indicator lamp
B_1 — 15V dc power source

CONCLUSIONS: In theory, this circuit is not new; only the application is different. You will remember that the emitter-base of a transistor must be forward biased to cause current in the collector circuit. Therefore, the BASE must be positive and EMITTER negative for the NPN transistor to cause collector current.

The timer, in its inoperative condition, is held in an OFF state because C is charged with the polarity indicated and the base of Q is negative. By pressing SW_1, two circuit actions happen. First, C discharges because SW_1 is a short circuit across C. Second, current flowing through R_1 makes the base end more positive and the transistor conducts. It will remain in conduction after SW_1 is released until C regains its charge and makes the base negative.

EXPERIENCE 4. Use the SECOND hand of your watch to measure the timing range of the circuit. Take readings when R_2 is at minimum and maximum resistance. Now, experiment with the circuit. Substitute other values for R_2 and C. Can you explain the effect on the circuit by changing these components? You may wish to review RC time constants in Unit 9.

CONCLUSIONS: Although a relay and lamp are used in this circuit, the relay could be used to switch any type of electric motor or power device.

PHOTO CONTROL CIRCUITS

The use of light-activated circuits further extends the science of electronics to control any type of machinery by varying the intensity of light. Photo devices are used to count products passing on an industrial conveyor. They are found in burglar alarm systems, safety and protective devices, camera light meters and thousands of other common and unusual circuits.

The conversion of light energy to electrical energy was discussed in Unit 4. This discussion covered the solar cell or sun cell. Also, the component which changed its resistance (photoresistive cell) was explained.

EXPERIENCE 5. Build an ELECTRONIC COUNTER. The basic circuit is illustrated in Fig. 12-34. To "turn on" the transistor and activate the relay and lamp, the BASE of the transistor must be POSITIVE. When a light beam falls on the photoresistive cell, it causes the cell to conduct. This current causes the BASE end of R_1 to become positive and the circuit will operate.

.

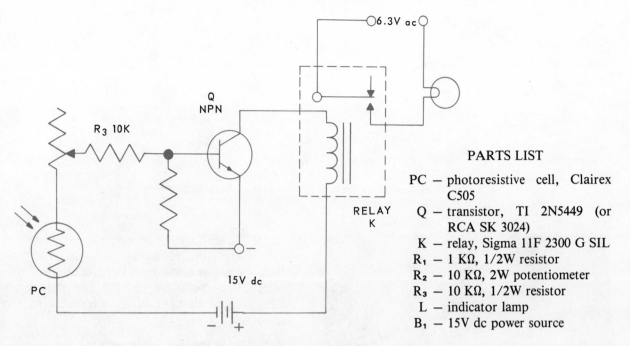

Fig. 12-34. The electronic counter circuit.

PARTS LIST

PC — photoresistive cell, Clairex C505
Q — transistor, TI 2N5449 (or RCA SK 3024)
K — relay, Sigma 11F 2300 G SIL
R_1 — 1 KΩ, 1/2W resistor
R_2 — 10 KΩ, 2W potentiometer
R_3 — 10 KΩ, 1/2W resistor
L — indicator lamp
B_1 — 15V dc power source

CONCLUSIONS: The purpose of the potentiometer R_2 is to vary total resistance of the base circuit and limit the current. Since the voltage at the base is the voltage drop across R_1 (E = IR), the R_2 will act as a SENSITIVITY CONTROL. If R_2 is set at a high resistance, then a brighter light must be used to decrease the resistance of the photocell.

If a numerical counter is attached in place of the indicator lamp, each time a light strikes the cell, it will be counted.

EXPERIENCE 6. Build a LIGHT OSCILLATOR. Change the circuit in Fig. 12-34 so that the indicating lamp is normally ON. In this connection, a light shining on the cell turns the indicator light OFF. Move the photocell near the indicator light. Now it will go crazy with flashing.

CONCLUSIONS: Turning the indicator light ON turns the transistor ON, which turns the light OFF, which turns the transistor OFF and the light ON. This wild circuit cannot decide just what it should do. Slight adjustments on R_2 may be required to make the circuit oscillate.

EXPERIENCE 7. Build a useful BURGLAR ALARM SYSTEM. The circuit combines the audio oscillator which you recently constructed in Unit 11 with a light control circuit to turn the audio tone ON or OFF, depending on connections. The block diagram and relay connections are illustrated in Fig. 12-35. Team up with a friend. One of you may build the control; the other builds the oscillator. Many modern burglar alarms may use ultrasonic waves to detect intruders. See Fig. 12-36.

CONCLUSIONS: It appears that a change in light intensity may not only be measured but can be made to turn lights or machines ON or OFF as desired.

EXPERIENCE 8. MEASURE TURBIDITY. A turbid liquid is like muddy water. It is cloudy. Usually, the turbidity of a liquid is the result of chemicals, dirt and solids being suspended in the liquid. Measuring turbidity is an important industrial test for water used in many processes. We will test turbidity by measuring the ability of a liquid to pass a beam of light.

Build the simple circuit shown in Fig. 12-37. Shine a flashlight through the glass jar of water onto the photocell. Read the conduction current of the transistor. Record this meter reading as a reference. Then, measure the turbidity of other liquids such as muddy water, milk and oil.

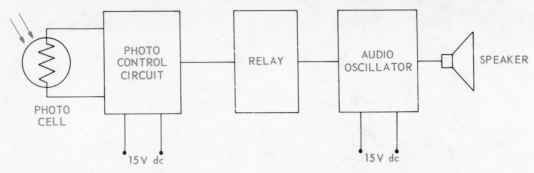

Fig. 12-35. Block diagram for the burglar alarm system. Separate power sources will be required if NPN transistors are used.

Fig. 12-36. Ultrasonic burglar alarm.   (Heath Co.)

CONCLUSIONS: The meter is truly reading in milliamperes. However, with a calibrated scale, the meter could read in percentage of turbidity. Very sensitive circuits can detect extremely light changes in the clearness of the liquid.

TEMPERATURE MEASUREMENT

Temperatures have been measured for years in industry, science and at home. The use of electronic circuitry and amplifiers not only has increased the accuracy of measurement, but now permits the scientist to observe extremely small changes in temperature with the human eye. Changes that can be measured and used to control other devices. A common example is in kitchen ovens which usually have automatic heat control.

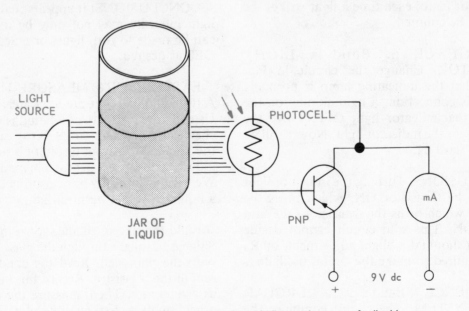

Fig. 12-37. This circuit measures the turbidity or clearness of a liquid.

Basically, a temperature measuring circuit will have a sensing device, an amplifier and a readout meter. In many industrial control systems, the readout is compared to some standard. Then, any difference is fed back to the heat source to correct its operation. See a typical digital thermometer in Fig. 12-38.

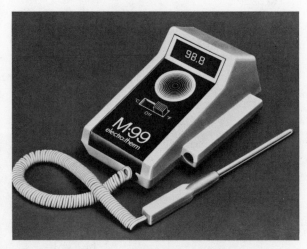

Fig. 12-38. Modern digital thermometer for humans. (Heath Co.)

EXPERIENCE 9. THERMISTORS. Build the simple thermistor circuit shown in Fig. 12-39. A THERMISTOR is a solid state component which changes its resistance in proportion to temperature. In the thermistor shown in Fig. 12-39, a rise in temperature decreases its resistance and there is an increase in current. The meter can be calibrated to read in degrees rather than current.

SMOKE DETECTORS

Fire detection systems used in homes and industry generally can be classified as ionization detectors and photoelectric detectors. Ionization detectors provide the earliest warning to fire. See Fig. 12-40. These devices use a radioactive energy source, usually Americum 241, that changes circulating air into a conductor of electric current. When smoke enters the ionization chamber, it causes an increase in current flow which sets off an alarm buzzer. Ionization detectors usually operate from a small battery.

Photoelectric detectors operate on the principle of reflected light. When smoke is visible in a chamber housing a photocell and light source, a reflection from the smoke is picked up by the photocell and the alarm is activated. This principle is similar to fog giving off light as automobile headlights shine on it.

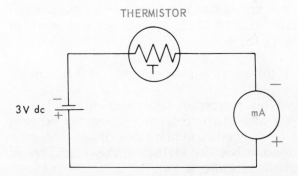

Fig. 12-39. In this thermistor circuit, a change of temperature produces a change of thermistor resistance and a change in current.

Fig. 12-40. Ionization smoke detector alarm. (Heath Co.)

ELECTRICAL ENERGY TO HEAT ENERGY

One of the most valuable effects of electricity is its ability to produce HEAT. When a current flows, it must overcome the resistance of the wire conductor and any other resistive components. You might compare this to mechanical friction. If you use your brakes too much while driving down a steep hill in your car, the brake drums and linings may become very hot and possibly burn.

In many cases, resistance is used in electronic circuits to limit current and the heat produced can be destructive. In this unit of study, however, you will purposely use resistance to produce HEAT.

RESISTANCE AND WATTS

In Unit 4, you found that POWER used by an electrical device is equal to the current times the voltage. Also, power is measured in WATTS:

P (in watts) = I (in amperes) × E (in volts)

While performing your experiences in Unit 4, you also discovered the relationship between volts, amperes and ohms in a circuit. This relationship is called OHM'S LAW which, stated mathematically, is:

$$I \text{ (in amperes)} = \frac{E \text{ (in volts)}}{R \text{ (in ohms)}}$$

or

I (in amperes) × R (in ohms) = E (in volts)

Now look back at the power formula. It also uses E in volts. Since E is equal to I × R, we will just put I × R in the formula in place of E. The new formula will then read:

P (in watts) = I in amperes × (I × R)

or

Power = $I^2 R$

This is one of the more important relationships in all electricity. POWER IN WATTS EQUALS THE CURRENT SQUARED TIMES RESISTANCE.

A neat little poem used in some of our Armed Services Schools will help you remember this formula:

"Twinkle, twinkle little star
Power equals I squared R"

NICHROME WIRE

A special wire called "Nichrome" is manufactured for use in applications where heat is re-

quired. It is an alloy of nickel and chromium and has approximately 50 times more resistance than copper wire of the same size.

Nichrome wire will become red hot if enough current flows through it. For this reason, Nichrome and similar kinds of wire are used as heating elements for stoves, toasters and heaters.

AN ELECTRIC STOVE

Assume that you wish to build a small electric stove which will use 550 watts of power when connected to your house electricity (117 volts). The current necessary for this power is:

P = I × E or 550 watts = I × 117 volts

I = 4.7 amperes

Next, compute the resistance which will permit this current to flow:

$$R = \frac{E}{I} \text{ or R (in ohms)} = \frac{117 \text{ volts}}{4.7 \text{ amperes}}$$

R = 25 ohms

Assume a Nichrome wire has a resistance of 2 ohms per foot. To get 25 ohms, a piece 12.5 ft. long is necessary. This length of wire is wound in a loose coil around some kind of insulating, nonburning material.

EXPERIENCE 10. Secure from your shop supply a piece of No. 24 Nichrome wire about two feet long. Wind it in a loose coil on some insulating ceramic material such as a glass rod, as illustrated in Fig. 12-41. Using the ohmmeter, measure and record the resistance of the Nichrome coil.

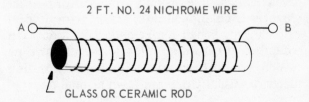

Fig. 12-41. A heating element is made by winding Nichrome wire on a ceramic rod.

Connect the resistance coil in the circuit displayed in Fig. 12-42.

Set voltage at one volt - observe current
Set voltage at two volts - observe current
Set voltage at five volts - observe current

Compute the power used at each voltage setting. CAUTION: THE RESISTANCE WIRE MAY GET RED HOT. DO NOT BURN YOURSELF OR START A FIRE.

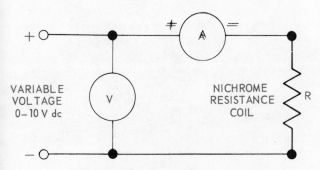

Fig. 12-42. Nichrome heating element is connected in a circuit. Voltage and current can be observed.

CONCLUSIONS: As the voltage is increased, the current also increases. The power increases as the square of the current. This means that if the current increases two times, the power increases two squared times or four times. Using a fixed voltage as found in your home, the power used by any light or appliance depends on the resistance of the light or appliance.

MICROWAVE COOKING

MICROWAVE cooking began at the Raytheon Company when, in 1945, Percy L. Spencer accidently melted a chocolate bar in front of a radar vacuum tube he was testing. Intrigued, Spencer sent out for popcorn and placed it in front of the tube. Kernels began popping, and a new concept for cooking was born.

Raytheon engineers began working immediately to develop the concept for the commercial food industry and, two years later, the first microwave oven was introduced.

The first microwave ovens were huge, weighing 750 pounds, and standing more that five feet tall. They also used a bulky, inefficient magnetron tube that was water cooled and required plumbing connections.

The first countertop microwave oven was introduced in 1967, and its speed and energy efficiency were well received in the consumer market. Fig. 12-43 shows the basic operation of a microwave oven. The magnetron tube produces the high frequency waves and they travel down a "waveguide" to be distributed by a "stirrer fan." The microwave energy penetrates the food and vibrates the molecules inside the food to produce heat and cooking results. See a modern microwave oven being used by a homemaker in Fig. 12-44.

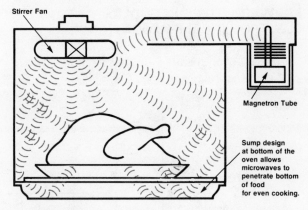

Fig. 12-43. Basic microwave operation.   (Amana)

Fig. 12-44. A modern microwave oven.   (Amana)

IC RADIO PROJECT

Fig. 12-45 shows a novel integrated circuit (IC) radio. The circuit uses a single, LM 3909 integrated circuit which acts as a detector amplifier. The tuning ability of this set is only as good as a "crystal radio," but a local radio station can provide listenable volume using a loudspeaker. Very low power drain allows a month of continuous radio operation from a single "D" flashlight cell.

The schematic for the IC radio is shown in Fig. 12-46. Note the high impedance, 40-45 ohm loudspeaker. The parts list is also given. Note that the antenna can be short (10 to 20 feet) for strong local reception or long (30 to 100 feet) for better overall reception. Happy listening!

FORWARD STEPS IN UNDERSTANDING ELECTRONIC COMMUNICATIONS

1. Oscillation means swinging back and forth or to vibrate.

2. An oscillator produces an alternating wave which continuously varies through a cycle of values.
3. The peak voltage of a wave is its maximum instantaneous voltage.
4. The frequency of a wave is the number of cycles of variation per second. It is measured in HERTZ.

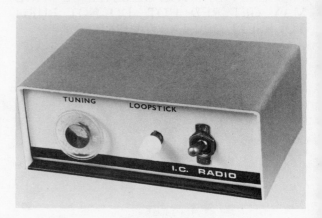

Fig. 12-45. Integrated Circuit (IC) Radio.

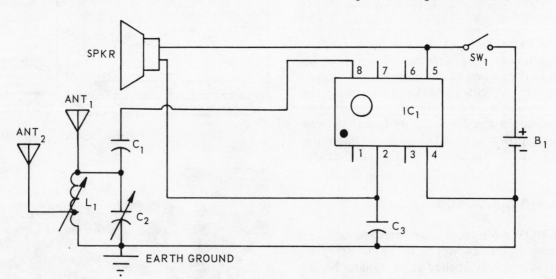

PARTS LIST FOR IC RADIO

C_1 — .001 μF capacitor, 50 WVdc
C_2 — 10-365 pF variable capacitor
C_3 — .1 μF capacitor, 50 WVdc
IC_1 — integrated circuit, National Semiconductor LM 3909
B_1 — 1 1/2V "D" cell
L_1 — loopstick antenna coil, with tapped coil

SW_1 — SPST switch
ANT_1 — short antenna, 10-20 ft. stranded antenna wire
ANT_2 — long antenna, 30-100 ft. stranded antenna wire
SPKR — 40-45 Ω speaker
Misc. — printed circuit materials, wire, solder, decals, IC socket, case

Fig. 12-46. Schematic and parts list for IC Radio.   (National Semiconductor Corp.)

5. The period of a wave is the time in seconds for one cycle.
6. The effective value of an ac wave is that value which produces the same heating power as a dc current of that value. It is equal to .707 × peak value.
7. A tank circuit is used for tuning. Its frequency depends on the values of L and C.
8. Interrupted continuous waves can be used for communication by keying to the Morse code.
9. When an audio tone is superimposed on a RF wave, it is called modulation.
10. Removing the audio tone from a RF signal is called demodulation or detection.
11. In AM communications, the amplitude of the RF signal is varied at an audio frequency rate.
12. In FM communications, the frequency of the RF wave is varied at an audio frequency rate.
13. An AM detector is sensitive to amplitude variations in the modulated signal.
14. An FM detector is sensitive to frequency variations in the modulated signal.
15. In America, the TV picture is produced by using 525 lines of scanning.
16. The line frequency in TV is 15,750 lines per second.
17. In TV, 30 complete pictures of even and odd lines are sent each second.
18. The electron beam in a TV tube is moved back and forth by magnetic deflection coils.
19. TV transmitter and TV receiver are kept in step and exact sweep frequency by sync pulses sent from the transmitter.
20. The sound for TV is transmitted by FM.
21. Color television uses separate electron guns for red, green and blue.
22. Color phosphors on a TV screen are arranged in triangular trios of red, green and blue.
23. A color TV transmitter uses the luminance and chrominance signals for modulation.
24. Your color television receiver has adjustments for hue and saturation. Hue refers to the color of some object. Saturation means richness of color.
25. The Shannon Communications Model includes the following elements: message source and message, coder, signal channel, decoder, message destination and noise.

FORWARD STEPS IN UNDERSTANDING ELECTRONICS IN MANUFACTURING INDUSTRIES

1. A motor converts electrical energy into mechanical energy.
2. The movement of a current-carrying conductor in a magnetic field is MOTOR ACTION.
3. A dc motor uses a commutator to change the direction of current in its armature.
4. An ac induction motor has stator windings and a rotor which follows the rotating field of the stator.
5. A rotating ac field is produced by out-of-phase currents in the stator coils.
6. A motor develops counter emf when it starts to rotate.
7. RPM can be measured by counting the pulses of voltage produced by a fixed magnet on the rotating wheel as it passes a pickup coil.
8. Electronic timing circuits use the time constant of a resistance-capacitance circuit to set the time interval.
9. Light intensity can be sensed by photo voltaic and photoresistive cells. The output of these cells can be used to control machinery.
10. A production line counter may use a photoelectric device.
11. Photoelectric devices can be used in counting and sorting machines, safety devices for machinery, burglar alarm systems and hundreds of other applications.
12. Photoelectric devices can be used to measure the turbidity of a liquid.
13. Thermistors change resistance in response to heat and can be used to measure temperature and control production processes.
14. Thermistors can be used to offset the effect of heat by changing the resistance of a circuit.
15. Resistance converts electrical energy to heat energy.
16. Power equals $I \times E$.
17. Power equals $I^2 R$.
18. Fire detection systems are classified as either ionization or photoelectric units. Ionization devices use a radioactive energy source. Photoelectric units operate on the principle of reflected light.

TEST YOUR KNOWLEDGE - UNIT 12

1. Build the TUNING CIRCUIT AND DIODE DETECTOR in Fig. 12-47. Connect earphones to the output. An antenna wire about 10 feet long may be required. The antenna coil may be the commercial variety or about 40 turns of No. 20 wire wound on a tube. How many stations can you tune in? While listening to a station, remove C_2 from the circuit. Is the sound louder or weaker?

2. Build a ONE TRANSISTOR RADIO. The circuit and parts list is supplied in Fig. 12-48. What is the purpose of the transistor in this circuit? Which components are used to tune in a station? What kind of a transistor is used? What is the purpose of C_2? How is detection accomplished?

3. Build a RADIO DETECTOR AND AMPLIFIER. In this problem, only the BLOCK DIAGRAMS will be supplied. Remember, you studied about transistor amplifiers in an early chapter? Build the system as described in Fig. 12-49. Note that ground connections are assumed to be there. Why do you think they are omitted? Which component causes demodulation? What kind of transistors are used? Can circuit be tuned by moving core in L_1?

4. The following appliances, operating on 117 volts ac, are rated by their power. What is the approximate resistance of each? (Do not write your answers in the book.)

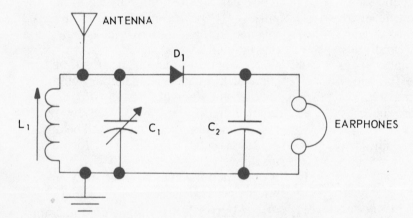

PARTS LIST

L_1 — Ferri loopstick or antenna coil
C_1 — 365 pF variable capacitor
C_2 — 100 pF ceramic capacitor
D_1 — crystal diode, 1N34
EAR — 2000 Ω crystal earphones

Fig. 12-47. A crystal detector.

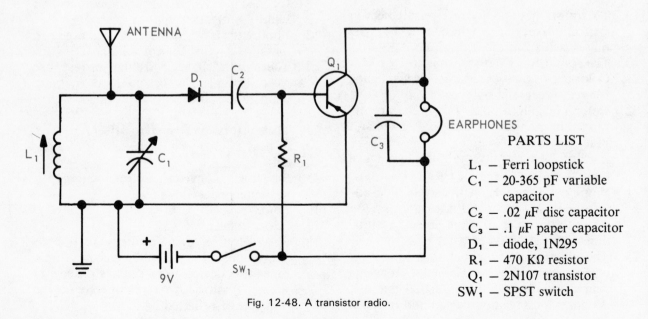

PARTS LIST

L_1 — Ferri loopstick
C_1 — 20-365 pF variable capacitor
C_2 — .02 μF disc capacitor
C_3 — .1 μF paper capacitor
D_1 — diode, 1N295
R_1 — 470 KΩ resistor
Q_1 — 2N107 transistor
SW$_1$ — SPST switch

Fig. 12-48. A transistor radio.

APPLIANCE	WATTAGE	RESISTANCE
Light bulb	100	
Toaster	550	
Electric heater	1000	

5. Devise a circuit which will measure the ability of a wall surface to reflect light.

6. Devise a circuit which may be used to detect fire and sound an alarm.

7. Why does a motor get hot when used to turn a heavy load?

8. Devise a circuit which you might use to count students passing in the school hallway.

9. A photo enlarger requires 5, 10 and 15 second exposure times. Devise a circuit which will electronically control the light.

10. Design the project and circuit for a front door or garden lamp which will automatically turn on at sunset and turn off at daybreak. See Fig. 12-50.

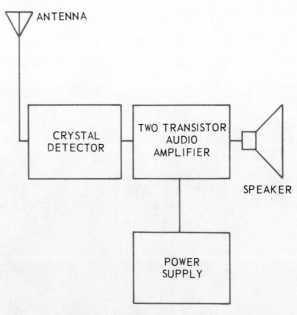

Fig. 12-49. The block diagram of detector and two transistor amplifier. Use shop power supply or transistor battery.

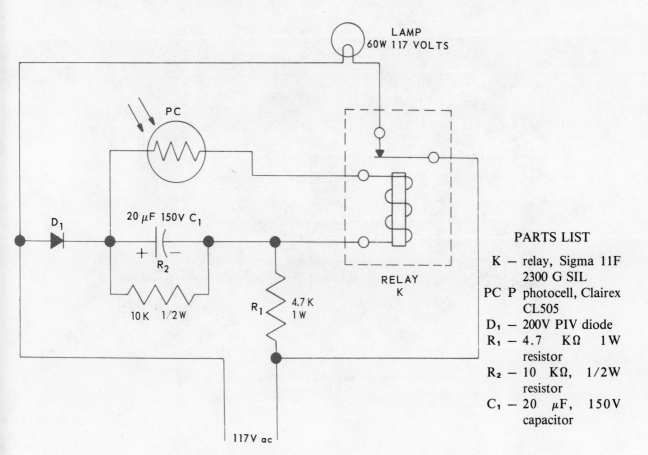

PARTS LIST

K — relay, Sigma 11F 2300 G SIL
PC P photocell, Clairex CL505
D_1 — 200V PIV diode
R_1 — 4.7 KΩ 1W resistor
R_2 — 10 KΩ, 1/2W resistor
C_1 — 20 μF, 150V capacitor

Fig. 12-50. A suggested circuit for porch or garden lamp control.

A one-chip computer developed for a variety of applications.
(Bell Laboratories)

REFERENCE SECTION

USING THE OSCILLOSCOPE

The oscilloscope is one of the basic instruments used for electronic measurement. In many ways, it works just like a TV set. However, instead of a complete picture, you will see a solid green line across the face of the Cathode Ray Tube (CRT). The line will vary up and down vertically, according to the signal being measured. Now, refer to Fig. A1-1 as we learn about the instrument controls.

VERTICAL POSITION. This knob moves the green pattern up or down so that you may center it on the screen.

HORIZONTAL POSITION. This control moves the pattern to right or left for centering on the face of the CRT.

SWEEP RANGE (Hz). This switch selects the frequency at which the spot on the face of the CRT moves from left to right.

V GAIN. Adjusts the weight of pattern up and down on screen face.

V ATT. A switch to set the oscilloscope to the range of peak-to-peak voltages to be measured.

SYNC. This control is used to adjust a pulse from external signal to lock-in or internal signal.

V INPUT and GND. These terminals will be used with test leads to apply signal to scope.

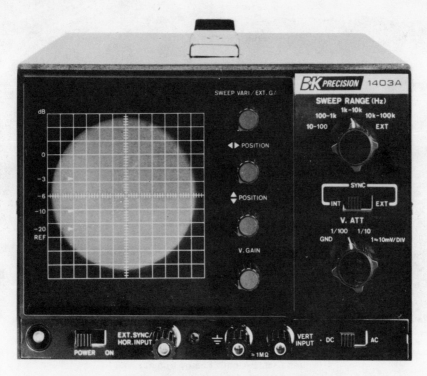

Fig. A1-1. A typical oscilloscope used in radio and TV service as well as in laboratory experiences.

Assume that you wish to observe and measure an ac signal with a frequency of 400 Hz and an amplitude of 20 volts peak-to-peak. Use audio generator.

1. Turn scope on and adjust intensity and focus for a clear, sharp, green line across the face. (Use only brightness necessary for good pattern. A very bright line shortens the life of the CRT.)
2. Center the green line using V and H positioning.
3. Adjust V GAIN until signal is displayed.
4. Set V ATT. to display signal.
5. Set SYNC. to INT.
6. Connect the signal to be measured to V IN— PUT and GND.
7. Set SWEEP to 100-1000 Hz range.
8. Adjust SWEEP VARI until two stationary sine wave patterns appear on the screen.

To calibrate scope so voltage peaks can be read directly, consult MANUAL for your particular oscilloscope. Practice using oscilloscope with other signals of different frequencies.

A specially adapted laser gun is being "fired" at a nearly invisible plume of affluents (pollutants) from the Tennessee Valley Authority's Bull Run Steam Plant to learn more about plume dispersal in the atmosphere. (United States Atomic Energy Commission)

USING THE AUDIO GENERATOR

You will use the audio generator, Fig. A2-1, in many of your laboratory experiences. It is used by technicians for servicing and adjusting equipment operating at frequencies up to 200 KHz. This instrument is a SIGNAL SOURCE.

OUTPUT TERMINALS. All signal outputs other than line frequency are taken from these terminals.

CONTROL AND DIAL. Use to set frequency of output signal. Use with,

FREQUENCY RANGE SWITCH. Dial reading times frequency range position will give output frequency. Example: Dial at 60. Range Switch at x2K. Frequency output, 60 x 2000 = 120,000 Hz.

ATTENUATOR. Determines amplitude of output wave in four steps (.01, .1, 1 and 10). Generator can produce either SINE waves or SQUARE waves.

OUTPUT. A control of the amplitude of the output wave within any range of attenuator.

EXPERIENCE: Connect audio generator directly to the V INPUT of the oscilloscope. Set generator at 400 Hz. Adjust oscilloscope to display four sine waves about one inch high in amplitude. Change frequency of generator and/or oscilloscope and make adjustments for a stable pattern. Consult oscilloscope manual to learn about CALIBRATION. Then, use the oscilloscope for direct measurement of generator output.

Fig. A2-1. A typical Audio Frequency Signal Generator which produces both sine and square waves up to 200 KHz. (RCA)

SIGNS AND SYMBOLS

BASIC CIRCUIT COMPONENTS

WIRING

WIRES CROSS, CONNECTED

WIRES CROSS, NOT CONNECTED

RESISTORS

RESISTOR FIXED VALUE

RESISTOR TAPPED

RESISTOR POTENTIOMETER

PHOTORESISTIVE CELL

CAPACITORS

CAPACITOR FIXED VALUE

CAPACITOR ELECTROLYTIC

CAPACITOR VARIABLE

GROUNDS

GROUND, EARTH

GROUND, CHASSIS

SOURCES

VOLTAIC CELL

BATTERY

PHOTOVOLTAIC CELL

THERMOCOUPLE

AC GENERATOR

METERS

VOLTMETER

AMMETER

OHMMETER

TRANSFORMERS/COILS

TRANSFORMER AIR CORE

TRANSFORMER IRON CORE

COILS, CHOKE AIR CORE

COIL, CHOKE IRON CORE

LAMPS

LAMP (INCANDESCENT)

LAMP, NEON

CIRCUIT PROTECTORS

FUSE

CIRCUIT BREAKER

BASIC CIRCUIT COMPONENTS (continued)

SWITCHES

SWITCH
SPST

SWITCH
SPDT

SWITCH
DPST

SWITCH
DPDT

PUSH BUTTON
Normally Open

PUSH BUTTON
Normally Closed

ROTARY
(MULTIPOSITION)

RELAYS

RELAY COIL

RELAY

RELAY,
DOUBLE CONTACT

AUDIO OUTPUT DEVICES

HEADPHONES

SPEAKER

CRYSTAL

CRYSTAL

CONNECTORS/PLUGS/JACKS

CONNECTOR,
PLUG IN

CONNECTOR,
RECEPTACLE

TERMINAL,
BINDING POST

JACK, PHONO

ANTENNA

ANTENNA

AUDIO INPUT DEVICES

PHONO PICKUP

MICROPHONE

SEMICONDUCTORS

DIODES

DIODE,
SEMICONDUCTOR

LIGHT EMITTING
DIODE (LED)

SCR
SILICON
CONTROLLED
RECTIFIER

ZENER DIODE

TRANSISTORS

TRANSISTOR,
PNP

TRANSISTOR,
NPN

INTEGRATED CIRCUIT (8 PIN)

GATES

AND

OR

NAND

NOR

INVERTER
(NOT)

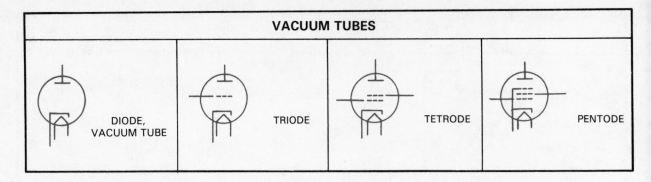

VACUUM TUBES

DIODE, VACUUM TUBE | TRIODE | TETRODE | PENTODE

GREEK LETTER SYMBOLS

Ω — Omega — ohms
μ — mu — amplification factor, micro
$\propto$ — alpha — current gain, CB transistor
β — beta — current gain, CE transistor
λ — lambda — wavelength
π — Pi — 3.1416
Δ — Delta — a change of

PREFIXES

pico	(p)	= 1/1000000 of 1/000000 of unit	1×10^{-12}
micro	(μ)	= 1/1000000 of the unit	1×10^{-6}
milli	(m)	= 1/1000 of the unit	1×10^{-3}
kilo	(K)	= 1000 units	1×10^{3}
mega	(M)	= 1000000 units	1×10^{6}

ABBREVIATIONS USED IN ELECTRONICS

R — RESISTANCE
L — INDUCTOR
C — CAPACITOR
X_L — INDUCTIVE REACTANCE
X_C — CAPACITIVE REACTANCE
Z — IMPEDANCE
E — VOLTAGE
V — VOLTAGE in Solid State Devices
Q or T — TRANSISTOR
V — VACUUM TUBE
CEMF — COUNTER ELECTROMOTIVE FORCE
A — GAIN
K — RELAY
RMS — ROOT MEAN SQUARE, Effective Value
T — TRANSFORMER
H — HENRY
F — FARAD
Hz — CYCLES PER SECOND
A — AMPERAGE
I — CURRENT
f — FREQUENCY
t — TIME
D — DIODE
LED — LIGHT EMITTING DIODE

CONVERSION CHART FOR PREFIXES

LARGER

10^6 . . . . MEGA (M)
0
0
0
10^3 . . . . KILO (K)
0
0
0
10^0 . . . . BASIC UNIT
0
0
10^{-3} . . . . MILLI (m)
0
0
0
10^{-6} . . . . MICRO (μ)
0
0
0
0
0
0
10^{-12} . . . . PICO (p)

SMALLER

EXAMPLES:

800 KHz = 800000 Hz
20 μF = .00002 F
100 pF = .0000000001 F
22 KΩ = 22000 Ω
1 MΩ = 1000000 Ω
10 mA = .01A

CONSTRUCTION MATERIALS

There is an old saying, "Anything worth doing is worth doing well." It is true that electronic circuits will often work if the component parts are placed helter-skelter and the interconnecting wires seem to resemble a bird's nest. However, this is not the mark of a good technician. Thought and order and neatness should show in any finished circuit or product. It is good advice to develop the habits of a skillful craftsworker early in your electronic construction experiences.

There are many kinds, types and varieties of hardware used in the assembly of electronic equipment.

BECOMING FAMILIAR WITH SIZES

A bolt or machine screw may be specified as:

6-32 x 1/2 RH MACHINE SCREW

The first number (6) refers to its size in diameter. The second number (32) means 32 threads per inch. The third figure (1/2) designates its length in inches. RH means Round Head, and Machine Screw tells you the type of screw.

SIZE	THREADS PER INCH	DIAMETER OF SCREW
2	56	.086 in.
4	36	.112 in.
4	40	.112 in.
6	32	.138 in.
8	32	.164 in.
10	32	.190 in.

Fig. A4-1. Screw sizes.

Fig. A4-1 shows typical sizes of screws, threads per inch and diameter in inches. This information will help you select the proper screw and matching nut as well as the size drill to use in making screw holes.

These sizes are mostly used in electronic equipment. The screws used should be nickel plated. Your projects should not be put together with iron stove bolts.

TYPES OF HARDWARE

Soldering lugs, Fig. A4-2, are used to make finished connections to parts and components. Wires may be crimped or soldered to the lug. These are sized by type, hole size and length.

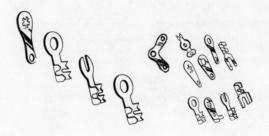

Fig. A4-2. Soldering lugs.   (G. C. Electronics)

Terminal strips are made from one lug to eight lugs and more. See Fig. A4-3. These are used to mount small components and to connect several wires to one point. The lug is bolted to the chassis.

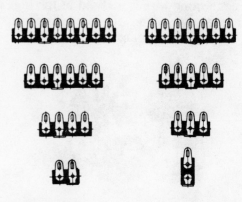

Fig. A4-3. Terminal strips.

Washers, Fig. A4-4, are used in many places. They are made in sizes to match machine screws. The shoulder washer is used with a flat fibre washer to insulate a binding post from a metal chassis.

Flat Metal Washer.

Lock Washer.

External Lock Washer.

Shoulder Fibre Washer.

Fig. A4-4. Washers.

Plugs and jacks are used on the ends of test leads and connectors when a quick easy disconnect is required. See Fig. A4-5.

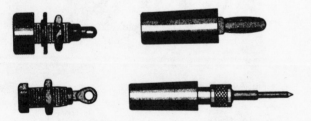

Fig. A4-5. Plugs and jacks.

Binding posts, Fig. A4-6, are used when it is necessary to connect wires to a circuit panel. The connecting wire is inserted in the hole and the top is screwed down. Some binding posts also have a banana plug jack in the top.

Fig. A4-6. Binding posts.

Battery and alligator clips are used on the ends of wires and test leads for making quick connections to circuits. See Fig. A4-7.

Fig. A4-7. Battery and alligator clips.

Screws are made in many sizes, lengths, head types and materials, Fig. A4-8. Nylon screws will be found on some electronic equipment.

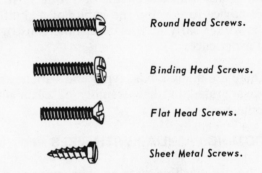

Round Head Screws.

Binding Head Screws.

Flat Head Screws.

Sheet Metal Screws.

Fig. A4-8. Bolt and screw heads.

Spade bolts are used to fasten components to panels and chassis. See Fig. A4-9.

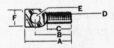

Fig. A4-9. Spade bolts.

Nuts are ordered by size and threads per inch; also by width and thickness. Hexagon nuts are shown in Fig. A4-10.

Fig. A4-10. Hexagon nuts.

Mounting nuts are used to mount controls, switches and other parts to the front panel of the equipment. See Fig. A4-11.

Fahnestock clips, Fig. A4-12, are used to make temporary connections of wires to a circuit or board.

Grommets are used to protect and insulate wires passing through a hole in a metal chassis. See Fig. A4-13.

Fig. A4-11. Mounting nuts.

Fig. A4-13. Grommets.

Fig. A4-12. Fahnestock clips.

Only a small sample of the hardware used in the assembly of electronic equipment is illustrated in this portion of the Reference Section. These common items will be needed in making simple projects.

ACKNOWLEDGMENTS

The efforts of many people were required to produce the original and revised editions of this text on exploration and discovery in electricity and electronics.
1. 1971 Edition.
   Thanks are given to Virtue B. Gerrish for editing and typing the original manuscript. Also, thanks to Perry Page, an industrial arts major at Humboldt State College, for preparing illustrations, building and checking experiences and activities.
2. 1981 Edition.
   Appreciation is expressed to Carrie Dugger for her support and understanding, and to Bill Dugger's children who make efforts such as this more rewarding. Many of the projects were constructed by Edward Dugger, son of Bill Dugger. Photos were done by VPI and SU Photo Lab. Margaret Dellapina and Toy Dugger typed the revised manuscript.
3. The authors wish to express their sincere thanks and appreciation to those in education and industry who generously contributed illustrations, product information and activities, without which this text could not have been written.

Amana, American Radio Relay League, AMP Incorporated, Ampex Corp., Atari Consumer Products Marketing, B & K Div. of Dynascan Corp., Bell Telephone Laboratories, Inc., Bernback Co., Boeing Aircraft Corp., British Airways, Bulova Watch Co., Bussmann Mfg. Co., Carolina Power and Light Co., CBC Industries, Centralab, Central Scientific Co., Clevite Corp., Burgess Battery Div., Cobra Div. of Dynascan Corp., Delco Products, General Motors Corp., Delco-Remy, General Motors Corp., Duracell International Inc., Eastman Kodak Co., EDUCOMP, Electric Energy Association, General Electric Co., Graymark Enterprises, Inc., Gulton Industries, Inc., Heath Co. (Heathkit), Hewlett-Packard, Inc., Howard W. Sams Co., Intel Corp., International Rectifier Corp., IRC, Inc., J.W. Miller Co., KHSL, Channel 12, Chico, CA, Kirsh Co., Lab-Volt, Buck Engineering Co., Lepel Corp., Lindberg Sola-Basic Industries, LTV — University Sound, Minneapolis Honeywell Regulator Co., Motorola Communications & Electronics Inc., NASA, National Semiconductor Corp., Office of European Atomic Energy Community, Ohmite Mfg. Co., Pacific Gas and Electric Co., Pacific Telephone Co., Philco-Ford Corp., Popular Electronics, Processor Technology Corp., Radar Devices Mfg. Corp., Radio Corporation of America (RCA), Radio Shack Corp., Rathcon, Inc., Sinclair Electronics, Sprague Products Co., Square D Co., Triad Transformer Corp., Triplett Corp., Union Carbide Corp., Union Switch and Signal Co., United Telecommunications Inc., U. S. Atomic Energy Commission, U. S. Department of Energy, Westinghouse Electric Corp., Viz Mfg. Co., Mr. Bob Fikes, Electronics Instructor in Alta Loma High School, CA

DICTIONARY OF TERMS

AC: Alternating current.

AC GENERATOR: Generator using slip rings and brushes to connect armature to external circuit. Output is alternating current.

ACTIVE DEVICE: A device used in electronic circuits which has the ability to change its state or control one type of signal with another one.

AGC: Abbreviation for AUTOMATIC GAIN CONTROL. Circuit employed to vary gain of amplifier in proportion to input signal strength so output remains at constant level.

AIR-CORE INDUCTOR: Inductor wound on insulated form without metallic core. Self-supporting coil without core.

ALIGNMENT: The adjustment of tuned circuits in amplifier and/or oscillator circuits so that they will produce a specified response at a given frequency.

ALKALINE CELLS: A primary cell that uses a manganese dioxide cathode, a zinc anode and an alkaline electrolyte.

ALNICO: Special alloy used to make small permanent magnets.

ALPHA: Greek letter ∞, represents current gain of a transistor. It is equal to change in collector current caused by change in emitter current for constant collector voltage.

ALTERNATING CURRENT (ac): Current of electrons that moves first in one direction and then in the other.

ALTERNATOR: An ac generator.

AMMETER: Meter used to measure current.

AMPERE: This unit measures electricity "on the move" or flowing in a circuit. Moving electricity is called current. It is this movement of electrical energy which does the work. It produces heat for your electric stove and light for your home. It produces music from your Hi-Fi Stereo and pictures and sound from television. Scientists have agreed that if a certain quantity of electricity passes a given point in a circuit in one second, it will be called one ampere. Smaller units than an ampere are:

MILLIAMPERE - one thousandth of an ampere or .001 ampere.

MICROAMPERE - one millionth of an ampere or .000001 ampere.

AMPERE-HOUR: Capacity rating measurement of batteries. A 100 ampere-hour battery will produce, theoretically, 100 amperes for one hour.

AMPERE TURN (IN): Unit of measurement of magnetomotive force. Represents product of amperes times number of turns in coil of electromagnet: mmf = 1.257 IN.

AMPLIFICATION: Ability to control a relatively large force by a small force. In a vacuum tube, relatively small variation in grid input signal is accompanied by relatively large variation in output signal.

AMPLIFICATION FACTOR: Expressed as μ (mu). Characteristic of vacuum tube to amplify a voltage. Mu is equal to change in plate voltage as result of change in grid voltage while plate current is constant.

AMPLIFIERS: POWER. Electron tube used to increase power output. Sometimes called a current amplifier.

VOLTAGE. An electron tube used to amplify a voltage.

AF AMPLIFIER. Used to amplify audio frequencies.

IF AMPLIFIER. Used to amplify intermediate frequencies.

RF AMPLIFIER. Used to amplify radio frequencies.

AMPLITUDE: Extreme range of varying quantity. Size, height of.

AMPLITUDE MODULATION (AM): Modulating a transmitter by varying strength of rf carrier at audio rate.

ANALOG: Data or information that is continuous over a range. Analog means the same as LINEAR.

AND CIRCUIT: A circuit with two or more inputs and all inputs must be present to produce an output signal.

ANODE: Positive terminal, such as plate in electron tube.

ANTENNA: Device for radiating or receiving radio waves.

ARITHMETIC AND LOGIC UNIT (ALU): Those circuits in a computer that perform arithmetic and logic functions on data.

ARMATURE: Revolving part in generator or motor. Vibrating or moving part of relay or buzzer.

ATOM: Smallest particle that makes up a type of material called an element.

ATOMIC NUMBER: Number of protons in nucleus of a given atom.

ATOMIC WEIGHT: Mass of nucleus of atom in reference to oxygen which has a weight of 16.

ATTENUATION: Decrease in amplitude or intensity.

AUTOMATION: An industrial technique of employing automatic, self-controlled machinery to replace human labor and control.

AVC: Automatic volume control.

AVERAGE VALUE: Value of alternating current or voltage of sine wave form that is found by dividing area under one alternation by distance along X axis between 0 and 180 deg. $E_{avg} = .637 E_{max}$.

AWG: AMERICAN WIRE GAUGE - used in sizing wire by numbers.

BAND: Group of adjacent frequencies in frequency spectrum.

BAND SWITCHING: Receiver employing switch to change frequency range of reception.

BANDWIDTH: Band of frequencies allowed for transmitting modulated signal.

BASE: The thin section between the emitter and collector of a transistor.

BATTERY: Several voltaic cells connected in series or parallel. Usually contained in one case.

BETA: Greek letter β, represents current gain of common-emitter connected transistor. It is equal to ratio of change in collector current to change in base current, while collector voltage is constant.

BIAS, FORWARD: Connection of potential to produce current across PN junction. Source potential connected so it opposes potential hill and reduces it.

BIAS, REVERSE: Connection of potential so little or no current will flow across PN junction. Source potential has same polarity as potential hill and adds to it.

BINARY: Number system having base of 2, using only the symbols 0 and 1.

BLEEDER: Resistor connected across power supply to discharge filter capacitors.

BRIDGE CIRCUIT: Circuit with series-parallel groups of components that are connected by common bridge. Bridge is frequently a meter in measuring devices.

BRIDGE RECTIFIER: Full-wave rectifier circuit employing four rectifiers in bridge configuration.

BRIDGE, WHEATSTONE: A bridge circuit for determining the value of an unknown component by comparison to one of known value.

BRUSH: Sliding contact, usually carbon, between commutator and external circuit in dc generator.

B-SUPPLY: Voltage supplied for plate circuits of electron tubes.

BYPASS CAPACITOR: Fixed capacitor which bypasses unwanted ac to ground.

CABLE: May be stranded conductor or group of single conductors insulated from each other.

CAPACITANCE: Inherent property of electric circuit that opposes change in voltage. Property of circuit whereby energy may be stored in electrostatic field.

CAPACITIVE REACTANCE (X_C): The opposition to the flow of an alternating current as the result of counter voltages stored in the capacitor when a varying voltage is applied. It is measured in OHMS and is inversely proportional to the frequency of applied voltage and the value of the capacitor in farads.

CAPACITOR: Device which possesses capacitance. Simple capacitor consists of two metal plates separated by insulator.

CAPACITY: Ability of battery to produce current over given length of time. Capacity is measured in ampere-hours.

CARRIER: Usually radio frequency continuous wave to which modulation is applied. Frequency of transmitting station.

CARRIER (in a semiconductor): Conducting hole or a electron.

CASCADE: Arrangement of amplifiers where output of one stage becomes input of next, throughout series of stages.

CATHODE: Emitter in electron tube.

CATHODE RAY TUBE: Vacuum tube in which electrons emitted from cathode are shaped into narrow beam and accelerated to high velocity before striking phosphor-coated viewing screen.

CENTER FREQUENCY: Frequency of transmitted carrier wave in FM when no modulation is applied.

CENTER TAP: Connection made to center of coil.

CHANNEL: The link or medium over which a transmitted message is sent.

CHOKE COIL: A high inductance coil used to prevent the passage of pulsating currents, but allows dc to pass.

CHOKE, RF: A choke coil with a high impedance at radio frequencies.

CIRCUIT BREAKER: Safety device which automatically opens circuit if overloaded.

CIRCUIT, ETCHED: The method of circuit board production in which the actual conduction paths on a copper-clad insulation board are coated with an acid resist. The board is then placed in an acid bath and unprotected parts of the copper-clad are eaten away, leaving the circuit conductors. Components are mounted and soldered between the conductors to form the completed circuit.

CIRCUIT, PRINTED: The method of printing circuit conductors on an insulated base. Component parts may also be printed or actual components soldered in place.

CITIZENS BAND: A band of frequencies allotted to two-way radio communications by private citizens. Operators are not required to pass technical examinations.

CODER: In a communications process, the coder accepts the message from the source and changes this message into some other form of transmission.

COLLECTOR: In a transistor, the semiconductor section which collects the majority carriers. Similar to the plate in a vacuum tube.

COMMON BASE: Transistor circuit, in which base is common to input and output circuits.

COMMON COLLECTOR: Transistor circuit in which collector is common to input and output circuits.

COMMON EMITTER: Transistor circuit in which emitter is common to input and output circuits.

COMMUTATION: The process of changing the alternating current in a generator armature into direct current in the external circuit by a mechanical switch consisting of commutator bars and brushes.

COMMUTATOR: Group of bars providing connections between armature coils and brushes. Mechanical switch to maintain current in one direction in external circuit.

COMPUTER: ANALOG. A computer which substitutes for any given physical quantity, a mechanical, electrical thermodynamic equivalent quantity that follows in direct proportion, the same laws of behavior as the original quantity. In general, the analog computer gives a continuous solution of the problem. Example: Slide rule, automobile speedometer.

DIGITAL. A computer which makes a one-to-one comparison or individual count to calculate. It solves problems in discrete steps, forming a discontinuous solution.

CONDUCTANCE (symbol G): Ability of circuit to conduct current. It is equal to amperes per volt and is measured in siemens. $G = \dfrac{1}{R}$

CONDUCTIVITY: The ability of a material to conduct an electric current. It is the reciprocal of resistivity.

CONDUCTIVITY, N TYPE: Conduction by electrons in N-type crystal.

CONDUCTIVITY, P TYPE: Conduction by holes in a P-type crystal.

CONDUCTOR: Material which permits free motion of large number of electrons.

CONTINUOUS WAVE (CW): Uninterrupted sinusoidal rf wave radiated into space, with all wave peaks equal in amplitude and evenly spaced along time axis.

CONTROL GRID: Grid in vacuum tube closest to cathode. Grid to which input signal is fed to tube.

CONTROL UNIT: In a computer, this unit interprets or decodes the program and produces the necessary signals to make the arithmetic unit perform its function.

COPPER LOSSES: Heat losses in motors, generators and transformers as result of resistance of wire.

CORE IRON: Magnetic materials usually in sheet form, used to form laminated cores for electromagnets and transformers.

COULOMB: A specified quantity of electrons. It is the amount of electricity moving in a circuit in one second by a current of one ampere. A Coulomb, designated by the letter Q, equals a charge of 6.24×10^{18} electrons.

COUNTER EMF (CEMF): Voltage induced in conductor moving through magnetic field which opposes source voltage.

COVALENT BOND: Atoms joined together, sharing each other's electrons to form stable molecule.

CROSSOVER NETWORK: The network designed to divide audio frequencies into bands for distribution to speakers.

CRYOGENICS: The use of electronic circuits designed to take advantage of increased efficiency at extremely low temperatures.

CRYSTAL DIODE: Diode formed by small semiconductor crystal and cat whisker.

CRYSTAL LATTICE: Structure of material when outer electrons are joined in covalent bond.

CURRENT: Transfer of electrical energy in conductor by means of electrons moving constantly and changing positions in vibrating manner.

CW: Abbreviation for Continuous Wave.

CYBERNETICS: The study of complex electronic computer systems and their relationship to the human brain.

CYCLE: A chain of events occurring in sequence. A voltage might start at zero and rise to a peak value in a positive direction, return to zero and then rise to peak value in a negative direction and then return to zero. One cycle is complete. Usually cycles continue to repeat this sequence of events.

D'ARSONVAL METER: Stationary-magnet moving coil meter.

DC: Direct current.

DC GENERATOR: Generator with connections to armature through a commutator. Output is direct current.

DCWV: Abbreviation for DIRECT CURRENT WORKING VOLTAGE. It is a specification of a capacitor.

DECAY: Term used to express gradual decrease in values of current and voltage.

DECAY TIME: The time required for a capacitor to discharge to a specified percentage of its original charge.

DECIBEL: One-tenth of a Bel. A unit used to express the relative increase or decrease in power. Unit used to express gain or loss in a circuit.

DECODER: The decoder receives the message, in a communications process, and reproduces it so that it can be understood at the message destination.

DEFLECTION: Deviation from zero of needle in meter. Movement or bending of an electron beam.

DEMODULATION: Process of removing modulating signal intelligence from carrier wave in radio receiver.

DETECTION: See DEMODULATION.

DETECTOR: CRYSTAL. A type of detection which uses the rectification characteristics of a crystal substance such as galena, silicon, germanium, iron pyrite. DIODE: A detector using a diode tube or semiconductor as the rectifier of the rf signal.

DIAPHRAGM: Thin disc, used in an earphone for producing sound.

DIELECTRIC: Insulating material between plates of capacitor.

DIGITAL: Data in discrete bits or pieces.

DIGITAL MULTIMETER: Meters that have light emitting diode displays or liquid crystal displays rather than having meter outputs.

DIODE: A two element vacuum tube or semiconductor. A diode conducts in the forward direction but offers high resistance to currents in the opposite direction.

DIODE DETECTOR: Detector circuit utilizing unilateral conduction characteristics of diode.

DIP SWITCH: A microminiature rocker switch.

DIRECT CURRENT (dc): Flow of electrons in one direction.

DOPING: Adding impurities to semiconductor material.

DRY CELL: Nonliquid cell, which is composed of zinc case, carbon positive electrode and paste of ground carbon, manganese dioxide and ammonium chloride as electrolyte.

DYNAMIC SPEAKER: Speaker which produces sound as result of reaction between fixed magnetic field and fluctuating field of voice coil.

EDISON EFFECT: Effect, first noticed by Thomas Edison, that emitted electrons were attracted to positive plate in vacuum tube.

EDDY CURRENTS: Induced current flowing in rotating core.

EDDY CURRENT LOSS: Heat loss resulting from eddy currents flowing through resistance of core.

EFFECTIVE VALUE: That value of alternating current of sine wave form that has equivalent heating effect of a direct current. ($.707 \times E_{peak}$)

EFFICIENCY: Ratio between output power and input power.

ELECTRODE: Elements in a cell.

ELECTRODYNAMIC SPEAKER: Dynamic speaker that uses electromagnetic fixed field.

ELECTROLYTE: Acid solution in a cell.

ELECTROLYTIC CAPACITOR: Capacitor with positive plate of aluminum. Dry paste or liquid forms negative plate. Dielectric is thin coat of oxide on aluminum plate.

ELECTROMAGNET: Coil wound on soft iron core. When current runs through coil, core becomes magnetized.

ELECTROMOTIVE FORCE (EMF): Force that causes free electrons to move in conductor. Unit of measurement is the volt.

Dictionary of Terms

ELECTRON: Negatively charged particle.

ELECTRON TUBE: Highly evacuated metal or glass shell which encloses several elements.

ELECTROSTATIC FIELD: Space around charged body in which its influence is felt.

ELEMENT: One of the distinct kinds of substances which either singly or in combination with other elements, makes up all matter in the universe.

EMISSION: Escape of electrons from a surface.

EMISSION, Types of: PHOTOELECTRIC. Emission of electrons as result of light striking surface of certain materials.

SECONDARY. Emission caused by impact of other electrons striking surface.

THERMIONIC: Process where heat produces energy for release of electrons from surface of emitter.

EMITTER: Element in a vacuum tube from which electrons are emitted. The CATHODE.

EMITTER: MAJORITY. In a transistor, the semiconductor section, either P or N type, which emits majority carriers into the interelectrode region.

MINORITY. In a transistor, the semiconductor section, either P or N type, which emits minority carriers into the interelectrode region.

ENERGY: That which is capable of producing work.

EPITAXY: The physical placement of materials on a semiconductor surface.

EXCITATION: To apply a signal to an amplifier circuit; to apply energy to an antenna system; to apply current to energize the field windings of a generator.

FARAD: This unit describes the capacity or size of a capacitor. It is the ability of a component to store an electric charge. It is agreed that a capacitor which stores a specified quantity of electrons, called a COULOMB, when a one volt electric force is applied to it, will have a value of one farad. The farad is a large unit. Smaller units used in electronics are:

MICROFARAD - one millionth of a farad or .000001 farad.

PICOFARAD - one millionth of one millionth of a farad or .000000000001 farad. (Used to be called a micromicrofarad.)

FEEDBACK: Transferring voltage from output of circuit back to its input.

FIELD MAGNETS: Electromagnets which make field of motor or generator.

FILTER: Circuit used to attenuate specific band or bands of frequencies.

FLUX: Represented by Greek letter Φ; total number of lines of magnetic force.

FLUX DENSITY (symbol B): Number of lines of flux per cross-sectional area of magnetic circuit.

FREQUENCY: The rate of number of times per second that a cycle occurs. Frequency is measured in "cycles per second" or HERTZ. These terms will also be used:

KILOHERTZ (KHz) = 1000 cycles per second.

MEGAHERTZ (MHz) = 1,000,000 cycles per second.

FREQUENCY BANDS: VLF - Very low frequencies, 10-30 KHz.

LF - Low frequencies, 30-300 KHz.

MF - Medium frequencies, 300-3000 KHz.

HF - High frequencies, 3-30 MHz.

VHF - Very high frequencies, 30-300 MHz.

UHF - Ultra high frequencies, 300-3000 MHz.

SHF - Super high frequencies, 3000-30,000 MHz.

EHF - Extremely high frequencies, 30,000-300,000 MHz.

FREQUENCY DEPARTURE: In FM, instantaneous change from center frequency of the audio modulating signal as a result of modulation.

FREQUENCY DEVIATION: Maximum departure from center frequency at the peak of the modulating signal.

FREQUENCY MODULATION (FM): Modulating transmitter by varying frequency of rf carrier wave at an audio rate.

FREQUENCY RESPONSE: Rating of device indicating its ability to operate over specified range of frequencies.

FREQUENCY SWING: The total frequency swing from maximum to minimum. It is equal to twice the deviation.

FULL-WAVE RECTIFIER: Rectifier circuit which produces a dc pulse output for each half cycle of applied alternating current.

FUNDAMENTAL: A sine wave that has the same frequency as complex periodic wave. Component tone of lowest pitch in complex tone. Reciprocal of period of wave.

FUSE: Safety protective device which opens an electric circuit if overloaded. Current above rating of fuse will melt fusible link and open circuit.

GAIN: Ratio of output ac voltage to input ac voltage.

GALVANOMETER: Meter which indicates very small amounts of current and voltage.

GATE: A circuit which permits an output only when a predetermined set of input conditions are met.

GAUSS: Measurement of flux density in lines per square centimeter.

GENERATOR: Rotating electric machine which provides a source of electrical energy. A generator converts mechanical energy to electrical energy.

GENERATORS, Types of: COMPOUND. Uses both series and shunt winding.

INDEPENDENTLY EXCITED. Field windings are excited by separate dc source.

SERIES. Field windings are connected in series with armature and load.

SHUNT. Field windings are connected across armature in shunt with load.

GERMANIUM: A rare grayish-white metallic chemical element. Symbol Ge; Atomic Wgt. 72.60; Atomic Number 32.

GILBERT: Unit of measurement of magnetomotive force. Represents force required to establish one maxwell in circuit with one Rel of reluctance.

GRID: A mesh of fine wire placed between cathode and plate of an electron tube.

GRID BIAS: Voltage between the grid and cathode, usually negative.

GRID DIP METER: A test instrument for measuring resonant frequencies, detecting harmonics and checking relative field strength of signals.

HALF-WAVE RECTIFIER: Rectifier which permits one-half of an alternating current cycle to pass and rejects reverse current of remaining half-cycle. Its output is pulsating dc.

HARMONIC FREQUENCY: Frequency which is multiple of fundamental frequency. Example: If fundamental frequency is 1000 KHz, then second harmonic is 2 × 1000 KHz or 2000 KHz; third harmonic is 3 × 1000 or 3000 KHz, and so on.

HEATER: Resistance heating element used to heat cathode in vacuum tube.

HEAT SINK: Mass of metal used to carry heat away from component.

HENRY: This unit describes the value of the inductance of a coil. If the current through a coil is made to vary, a voltage is produced by the coil. This is called INDUCTANCE. If the current is changing at the rate of one ampere per second and a voltage of one volt is induced, the coil is said to have an inductance of one henry. Smaller values are also used in electronics:

MILLIHENRY - one thousandth of a henry or .001 henry.

MICROHENRY - one millionth of a henry or .000001 henry.

HERTZ: The unit of measurement for frequency, named in honor of Heinrich Hertz who discovered radio waves. One HERTZ equals "one cycle per second."

HETERODYNE: The process of combining two signals of different frequencies to obtain the difference frequency.

HOLE: Positive charge. A space left by removed electron.

HOLE INJECTION: Creation of holes in semiconductor material by removal of electrons by strong electric field around point contact.

HORSEPOWER: 33,000 ft. lbs. of work per minute or 550 ft. lbs. of work per second equals one horse power. Also, 746 watts = 1 HP.

HUM: Form of distortion introduced in an amplifier as result of coupling to stray electromagnetic and electrostatic fields or insufficient filtering.

HYDROMETER: Bulb-type instrument used to measure specific gravity of a liquid.

HYSTERESIS LOSS: Energy loss in substance as molecules or domains move through cycle of magnetization. Loss due to molecular friction.

IMPEDANCE MATCHING: Sometimes called Z match; the matching of two different impedances to obtain maximum transfer of power.

IMPEDANCE (Z): The vector sum of the resistance, inductive reactance and the capacitive reactance in a circuit. R, X_C and X_L cannot be added directly since both the opposition in ohms of X_L and X_C are not in phase with R.

IMPURITY: Atoms within a crystalline solid which are foreign to the crystal.

IN PHASE: Two waves of the same frequency are in phase when they pass through their maximum and minimum values at the same instant with the same polarity.

INDUCED CURRENT: Current that flows as result of an induced EMF.

INDUCED EMF: Voltage induced in conductor as it moves through magnetic field.

INDUCTANCE: Inherent property of electric circuit that opposes a change in current. Property of circuit whereby energy may be stored in magnetic field.

INDUCTANCE, MUTUAL: See MUTUAL INDUCTANCE.

INDUCTION MOTOR: An ac motor operating on principle of rotating magnetic field produced by out-of-phase currents. Rotor has no electrical connections, but receives energy by transformer action from field windings. Motor torque is developed by interaction of rotor current and rotating field.

INDUCTIVE REACTANCE (X_L): the opposition to the flow of an alternating current as the result of counter voltages induced in the coil by a varying current. It is measured in OHMS and is directly proportional to the frequency of the applied voltage and the value of the coil in henrys.

INDUCTOR: A coil, a component with the properties of inductance.

INPUT IMPEDANCE: The impedance of the input terminals of a circuit or device, with the input generator disconnected.

INPUT/OUTPUT: That area of a computer that provides a means of communication with the inside of the computer.

INSTANTANEOUS VALUE: Any value between zero and maximum depending upon instant selected.

INSULATORS: Substances containing very few free electrons and requiring large amounts of energy to break electrons loose from influence of nucleus.

INTEGRATED CIRCUIT: A concentration of transistors, diodes, resistors and capacitors on a microminiature chip.

INTERNAL RESISTANCE: Refers to internal resistance of source of voltage or EMF. A battery or generator has internal resistance which may be represented as a resistor in series with source.

INTERRUPTED CONTINUOUS WAVE (ICW): Continuous wave radiated by keying transmitter into long and short pulses of energy (dashes and dots) conforming to code such as Morse Code.

INTERSTAGE: Existing between stages, such as an interstage transformer between two stages of amplifiers.

ION: An atom which has lost or gained some electrons. It may be positive or negative depending on the net charge.

IONIZATION: An atom is said to be ionized when it has lost or gained one or more electrons.

IONIZATION SMOKE DETECTOR: A smoke detector that operates from passing electric current through an ionization chamber.

IONOSPHERE: Atmospheric layer from 40 to 350 miles above the earth, containing a high number of positive and negative ions.

JUNCTION DIODE: PN junction having unidirectional current characteristics.

JUNCTION TRANSISTOR: Transistor consisting of thin layer of N or P type crystal between P or N type crystals. Designated as NPN or PNP.

KEY: Manually operated switch used to interrupt rf radiation of transmitter.

KEYING: Process of causing CW transmitter to radiate an rf signal when key contacts are closed.

KILO: Prefix meaning one thousand times.

KILOWATT-HOUR (KWH): Means 1000 watts per hour. Common unit of measurement of electrical energy for home and industrial use. Power is priced by the KWH.

LAMBDA: Greek letter λ. Symbol for wavelength.

LAMINATIONS: Thin sheets of steel used in cores of transformers, motors and generators.

Dictionary of Terms

LAWS OF MAGNETISM: Like poles repel; unlike poles attract.

LEAD ACID CELL: Secondary cell which uses lead peroxide and sponge lead for plates, and sulfuric acid and water for electrolyte.

LEFT-HAND RULE: A method, using your left hand, to determine polarity of an electromagnetic field or direction of electron flow.

LENZ'S LAW: Induced EMF in any circuit is always in such a direction as to oppose effect that produces it.

LIGHT EMITTING DIODE (LED): A PN junction that emits light when biased in the forward direction.

LINEAR: In a straight line; a mathematical relationship in which quantities vary in direct proportion.

LINEAR AMPLIFIER: An amplifier whose output is in exact proportion to its input.

LINEAR DEVICE: Electronic device or component whose current-voltage relation is a straight line.

LINES OF FORCE: Graphic representation of electrostatic and magnetic fields showing direction and intensity.

LOAD: Resistance connected across circuit which determines current flow and energy used.

LOADING A CIRCUIT: Effect of connecting voltmeter across circuit. Meter will draw current and effective resistance of circuit is lowered.

LOCAL OSCILLATOR: Oscillator in superheterodyne receiver, output of which is mixed with incoming signal to produce intermediate frequency.

LODESTONE: Natural magnet, so called a "leading stone" or lodestone because early navigators used it to determine directions.

LOGICAL FUNCTIONS: An expression referring to a definite state of condition.

AND FUNCTION. An output is obtained only with a combined group of input signals.

OR FUNCTION. An output is obtained with any one of a group of input signals.

NOT FUNCTION: An output is obtained only when there is no input signal.

MEMORY FUNCTION: An output is continually obtained unless an input signal is applied.

LOUDSPEAKER: Device to convert electrical energy into sound energy.

L PAD: A combination of two variable resistors, one in series and the other across the load, which is used to vary the output of an audio system and match impedances.

MAGNET: Substance that has the property of magnetism.

MAGNETIC CIRCUIT: Complete path through which magnetic lines of force may be established under influence of magnetizing force.

MAGNETIC FIELD: Imaginary lines along which magnetic force acts. These lines emanate from N pole and enter S pole, forming closed loops.

MAGNETIC FLUX (symbol Φ phi): entire quantity of magnetic lines surrounding a magnet.

MAGNETIC LINES OF FORCE: Magnetic line along which compass needle aligns itself.

MAGNETIC MATERIALS: Materials such as iron, steel, nickel and cobalt which are attracted to magnet.

MAGNETIC PICKUP: Phono cartridge which produces an electrical output from armature in magnetic field.

Armature is mechanically connected to reproducing stylus.

MAGNETIZING CURRENT: Current used in transformer to produce transformer core flux.

MAGNETOMOTIVE FORCE (F) (mmf): Force that produces flux in magnetic circuit.

MAGNET POLES: Points of maximum attraction on a magnet; designated as north and south poles.

MAJOR CARRIER: Conduction through semiconductor as a result of majority of electrons or holes.

MATTER: Physical substance of common experience. Everything about us is made up of matter.

MAXIMUM VALUE: Peak value of sine wave either in positive or negative direction.

MEGA: Prefix meaning one million times.

MEMORY: That part of the computer that stores data and programs.

MERCURY CELLS: A primary cell with a mercury oxide cathode, a zinc anode and an alkaline electrolyte.

MESSAGE SOURCE: That part of a communications system that contains the information to be communicated.

MHO: Unit of measurement of conductance (siemen).

MICA CAPACITOR: Capacitor made of metal foil plates separated by sheets of mica.

MICRO: Prefix meaning one millionth of.

MICROMICRO: Prefix meaning one millionth of one millionth of.

MICROPHONE: Energy converter that changes sound energy into corresponding electrical energy.

MIL: One thousandth of an inch. (.001 inch)

MILLI: Prefix meaning one thousandth of.

MILLIAMMETER: Meter which measures in milliampere range of currents.

MINOR CARRIER: Conduction through semiconductor opposite to major carrier. Example: if electron is major carrier, then hole is minor carrier.

MINUS (symbol −): Negative terminal of junction of circuit.

MODULATION: Process by which amplitude or frequency of sine wave voltage is made to vary according to variations of another voltage or current called modulation signal.

MODULATION PRODUCT: Sideband frequencies resulting from modulation of a radio wave.

MOLECULE: Smallest division of matter. If further subdivision is made, matter will lose its identity.

MOTOR: Device which converts electrical energy into mechanical energy.

MOTORS, Types of dc: COMPOUND. Uses both series and parallel field coils.

SERIES. Field coils are connected in series with armature circuit.

SHUNT. Field coils are connected in parallel with armature circuit.

MU: Greek letter (μ) used to represent the amplification factor of a vacuum tube; magnetic permeability; the prefix meaning one millionth of.

MULTIMETER: A combination volt, ampere and ohm-meter.

MULTIPLIER: Resistance connected in series with meter movement to increase its voltage range.

MULTIVIBRATORS, Types of: ASTABLE. A free-running multivibrator.

BISTABLE. A single trigger pulse switches conduction from one tube to the other.

FREE-RUNNING. Frequency of oscillation depending upon value of circuit components. Continuous oscillation.

MONOSTABLE. One trigger pulse is required to complete one cycle of operation.

ONE SHOT. Same as MONOSTABLE.

MUTUAL INDUCTANCE (M): When two coils are so located that magnetic flux of one coil will cause an EMF in the other, there is mutual inductance.

NATURAL MAGNET: Magnets found in natural state in form of mineral called magnetite.

NEGATIVE ION: Atom which has gained electrons and is negatively charged.

NETWORK: Two or more components connected in either series or parallel.

NEUTRON: Particle which is electrically neutral.

NICKEL CADMIUM CELL: Alkaline cell with paste electrolyte hermetically sealed.

NO LOAD VOLTAGE: Terminal voltage of battery or supply when no current is flowing in external circuit.

NONLINEAR DEVICE: Electronic device or component whose current-voltage relation is not a straight line.

NUCLEONICS: The branch of physics dealing with the science of small particles and the release of energy from the atom.

NUCLEUS: Core of the atom.

OHM: Electricity may or may not experience some difficulty or opposition as it flows through a circuit. This is called resistance and it is measured in ohms. Some materials have very high resistance; others allow electricity to flow easily. A circuit containing one ohm of resistance will allow one ampere of current to flow if one volt of electrical force is connected to it. Larger units of resistance may be written as:

KILOHM - one thousand ohms or 1000 ohms.

MEGOHM - one million ohms or 1,000,000 ohms.

OHMMETER: Meter used to measure resistance in ohms.

OHM'S LAW: Mathematical relationship between current, voltage and resistance discovered by George Simon Ohm.

$$I = \frac{E}{R} \qquad E = IR \qquad R = \frac{E}{I}$$

OHMS PER VOLT: Unit of measurement of sensitivity in a meter.

OR CIRCUIT: A circuit with two or more inputs and one input signal must be present to produce an output signal.

OSCILLATOR: An electron tube generator of alternating current voltages.

OSCILLATORS, Types of: ARMSTRONG. An oscillator using tickler coil for feedback.

COLPITTS. An oscillator using split tank capacitor as feedback circuit.

CRYSTAL-CONTROLLED. An oscillator controlled by piezoelectric effect.

ELECTRON COUPLED OSCILLATOR (ECO): Combination oscillator and power amplifier utilizing electron stream as coupling medium between grid and plate tank circuits.

HARTLEY. Oscillator using inductive coupling of tap-ped tank coil for feedback.

PUSH-PULL. Push-pull circuit utilizing interelectrode capacitance of each tube to feed back energy to grid circuit to sustain oscillations.

RC OSCILLATORS. Oscillators depending upon charge and discharge of capacitor in series with resistance.

OSCILLOSCOPE: Test instrument using cathode ray tube, permitting observation of signal.

OUTPUT: The energy from a circuit or device.

PARALLEL CIRCUIT: Circuit which contains two or more paths for electrons supplied by common voltage source.

PARALLEL RESONANCE: Parallel circuit of an inductor and capacitor at frequency when inductive and capacitive reactances are equal. Current in capacitive branch is 180 deg. out of phase with inductive current and their vector sum is zero.

PASSIVE DEVICE: An electrical device that does not have the ability to control its state.

PEAK: Maximum value of sine wave; the highest voltage current or power reached during a particular cycle or operating time.

PEAK INVERSE VOLTAGE: Value of voltage applied in reverse direction across diode.

PEAK INVERSE VOLTAGE RATING: The inverse voltage a diode will withstand without arcback.

PEAK-TO-PEAK: Measured value of sine wave from peak in positive direction to peak in negative direction.

PEAK VALUE: Maximum value of an alternating current or voltage.

PENTAGRID CONVERTER: Tube with five grids.

PENTAVALENT: Semiconductor impurity having five valence electrons. Donor impurities.

PENTODE: Electron tube with five elements including cathode, plate, control grid, screen grid and suppressor grid.

PERIPHERAL: Units of a computer that work in exterior (not a basic part of the computer).

PERIOD: The time in seconds elapsed during one cycle. It is equal to one divided by frequency, $P = \frac{1}{f}$

PERMANENT MAGNET: Bars of steel and other substances which have been permanently magnetized.

PERMEABILITY (symbol μ): Relative ability of substance to conduct magnetic lines of force as compared with air.

PHASE: Relationship between two vectors in respect to angular displacement.

PHASE SPLITTER: Amplifier which produces two waves that have exactly opposite polarities from single input wave form.

PHOTODIODE: A PN junction diode which conducts upon exposure to light energy.

PHOTOELECTRIC CELL: A cell which produces an electric potential when exposed to light.

PHOTOELECTRIC EFFECT: The property of certain substances to emit electrons when subjected to light.

PHOTOELECTRIC SMOKE DETECTOR: A smoke detector that operates from the electrical effort of the density of smoke passing through a reflected light area.

PHOTOELECTRONS: Electrons emitted as a result of light.

PHOTOMASK: A transparent glass plate carrying patterns for electronic active and passive devices.

PHOTON: A discrete quantity of electromagnetic energy; a quantum.

PHOTOSENSITIVE: Characteristic of material which emits electrons from its surface when energized by light.

PHOTOTUBE: Vacuum tube employing photo sensitive material as its emitter or cathode.

PHOTOVOLTAIC: The generation of a voltage at the junction of two materials when exposed to light.

PIEZOELECTRIC EFFECT: Property of certain crystalline substances of changing shape when an EMF is impressed upon crystal. Action is also reversible.

PITCH: Property of musical tone determined by its frequency.

PLANETARY ELECTRONS: Electrons considered in orbit around the nucleus of an atom.

PLATE: Anode of vacuum tube. Element in tube which attracts electrons.

PLUS (symbol +): Positive terminal or junction of circuit.

PM SPEAKER: Speaker employing permanent magnet as its field.

PN JUNCTION: The line of separation between N-type and P-type semiconductor materials.

POINT CONTACT DIODE: Diode consisting of point and a semiconductor crystal.

POLARITY: Property of device or circuit to have poles such as north and south or positive and negative.

POLARIZATION: Defect in cell caused by hydrogen bubbles surrounding positive electrode and effectively insulating it from chemical reaction.

POLES: Number of poles in motor or generator field.

POSITIVE ION: Atom which has lost electrons and is positively charged.

POT: Abbreviation for potentiometer.

POWER: Rate of doing work. In dc circuits, $P = I \times E$.

POWER AMPLIFICATION: Ratio of output power to input grid driving power.

POWER SUPPLY: Electronic circuit designed to provide various ac and dc voltages for equipment operation. Circuit may include transformers, rectifiers, filters and regulators.

POWER TRANSISTOR: Transistors designed to deliver a specified output power level.

PREAMPLIFIER: Sensitive low-level amplifier with sufficient output to drive standard amplifier.

PREFIXES: A word united or joined to the beginning of another word to change its meaning. The new approved multiples are shown in the accompanying chart.

PREFIXES: Multiples

10^{12}	tera-	10^{-1}	deci-
10^{9}	giga-	10^{-2}	centi-
10^{6}	mega-	10^{-3}	milli-
10^{4}	myria-	10^{-6}	micro-
10^{3}	kilo-	10^{-9}	nano-
10^{2}	hekto-	10^{-12}	pico-
10	deka-		

PRIMARY CELL: Cell that cannot be recharged.

PRIMARY WINDING: Coil of transformer which received energy from ac source.

PROBE: A pointed metal end of a test lead, to contact specific points in a circuit to be measured.

PROTON: Positively charged particle.

PULSE: Sudden rise and fall of a voltage or current.

SPIKE. An unwanted pulse of relatively short duration superimposed on a main pulse.

SYNC. A pulse sent by a TV transmitter to synchronize the scanning of the receiver with the transmitter; a pulse used to maintain predetermined speed and/or phase relations.

PUSH-PULL AMPLIFIER: Two tubes used to amplify signal in such a manner that each tube amplifies one half-cycle of signal. Tubes operate 180 deg. out of phase.

Q: Letter representation for quantity of electricity (coulomb).

Q: Quality, figure of merit; ratio between energy stored in inductor during time magnetic field is being established to losses during same time. $Q = \dfrac{X_L}{R}$

QUIESCENT: At rest. Inactive.

RADIO FREQUENCY CHOKE (RFC): Coil which has high impedance to rf currents.

RADIO SPECTRUM: Division of electromagnetic spectrum used for radio.

RADIO WAVE: A complex electrostatic and electromagnetic field radiated from a transmitter antenna.

REACTANCE: INDUCTIVE (X_L). The opposition to the flow of an alternating current as the result of counter voltages induced in the coil by a varying current. It is measured in OHMS and is directly proportional to the frequency of the applied voltage and the value of the coil in henrys.

CAPACITIVE (X_C). The opposition to the flow of an alternating current as the result of counter voltages stored in the capacitor when a varying voltage is applied. It is measured in OHMS and is inversely proportional to the frequency of applied voltage and the value of the capacitor in farads.

RECIPROCAL: Reciprocal of number is one divided by the number.

RECTIFICATION: The process of changing ac to pulsating dc.

RECTIFIER: Component or device used to convert ac into a pulsating dc.

REED RELAY: A relay which has two incapsulated metal strips mounted inside a coil of wire. When the coil is energized, the metal strips are either repelled or attracted.

REGULATION: Voltage change that takes place in output of generator or power supply when load is changed.

REGULATION, PERCENTAGE OF: Percentage of change in voltage from no-load to full-load in respect to full-load voltage. Expressed as:

$$\frac{E_{no\ load} - E_{full\ load}}{E_{full\ load}} \times 100$$

REJECT CIRCUIT: Parallel tuned circuit at resonance. Rejects signals at resonant frequency.

RELAY: Magnetic switch.

RELAXATION OSCILLATOR: Nonsinusoidal oscillator whose frequency depends upon time required to charge or discharge capacitor through resistor.

REPULSION-START MOTOR: A motor which develops starting torque by interaction of rotor currents and single-phase stator field.

RESIDUAL MAGNETISM: Magnetism remaining in material after magnetizing force is removed.

RESISTANCE: Quality of electric circuit that opposes flow of current through it.

RESONANT FREQUENCY: Frequency at which tuned circuit oscillates. (See TUNED CIRCUIT.)

RETENTIVITY: Ability of material to retain magnetism after magnetizing force is removed.

RETRACE: Process of returning scanning beam to starting point after one line is scanned.

RIPPLE VOLTAGE: An ac component of dc output of power supply due to insufficient filtering.

RMS VALUE: ROOT-MEAN-SQUARE value. The same as effective value ($.707 \times E_{peak}$)

ROTOR: Rotating part of an ac generator.

SAWTOOTH GENERATOR: Electron tube oscillator producing sawtooth wave.

SAWTOOTH WAVE: Wave shaped like the teeth of a saw.

SCHEMATIC: Diagram of electronic circuit showing electrical connections and identification of various components.

SCREEN GRID: Second grid in electron tube between grid and plate, to reduce interelectrode capacitance.

SECONDARY CELL: Cell that can be recharged by reversing chemical action with electric current.

SECONDARY EMISSION: Emission of electrons as result of electrons striking plate of electron tube.

SECONDARY WINDING: Coil which receives energy from primary winding by mutual induction and delivers energy to load.

SELECTIVITY: Relative ability of receiver to select desired signal while rejecting all others.

SELF-INDUCTANCE: EMF is self-induced when it is induced in conductor carrying current.

SEMICONDUCTOR: Conductor with resistivity somewhere in range between conductors and insulators.

SEMICONDUCTOR, N TYPE: Semiconductor which uses electrons as major carrier.

SEMICONDUCTOR, P TYPE: Semiconductor which uses holes as major carrier.

SENSITIVITY: Ability of circuit to respond to small signal voltages.

SENSITIVITY OF METER: Indication of loading effect of meter. Resistance of moving coil and multiplier divided by voltage for full scale deflection. Sensitivity equals one divided by current required for full scale deflection. Example: A 100 μA meter movement has sensitivity of $\dfrac{1}{.0001}$ or 10,000 ohms/volt.

SERIES CIRCUIT: Circuit which contains only one possible path for electrons through circuit.

SERIES PARALLEL: Groups of series cells with output terminals connected in parallel.

SERIES RESONANCE: Series circuit of inductor, a capacitor and resistor at a frequency when inductive and capacitive reactances are equal and cancelling. Circuit appears as pure resistance and has minimum impedance.

SHIELD: Partition or enclosure around components in circuit to minimize effects of stray magnetic and radio frequency fields.

SHORT CIRCUIT: Direct connection across source which provides zero resistance path for current.

SHUNT: To connect across or parallel with circuit or component.

SHUNT: Parallel resistor to conduct excess current around meter moving coil. Shunts are used to increase range of meter.

SIDEBANDS: Frequencies above and below carrier frequency as the result of modulation.

LOWER: Frequencies equal to difference between carrier and modulating frequencies.

UPPER: Frequencies equal to carrier plus modulating frequencies.

SIEMEN: Unit of measurement of conductance.

SIGNAL: The intelligence, message or effect to be sent over a communications system; an electrical wave corresponding to intelligence.

SINE WAVE: A wave form of a single frequency alternating current. A wave whose displacement is the sine of an angle proportional to time or distance.

SINGLE ENDED AMPLIFIER: An amplifier whose final power stage is a single vacuum tube or transistor.

SINGLE PHASE MOTOR: Motor which operates on single phase alternating current.

SLIP RINGS: Metal rings connected to rotating armature windings in generator. Brushes sliding on these rings provide connections for external circuit.

SOLENOID: Coil of wire carrying electric current possessing characteristics of magnet.

SPACE CHARGE: Cloud of electrons around cathode of an electron tube.

SPECIFIC GRAVITY: Weight of liquid in reference to water, which is assigned value of 1.0.

SPLIT PHASE MOTOR: Single phase induction motor which develops starting torque by phase displacement between field windings.

SQUIRREL CAGE ROTOR: Rotor used in an induction motor made of bars placed in slots of rotor core and all joined together at ends.

STABILITY: The ability to stay on a given frequency or in a given state without undesired variation.

STAGE: Section of an electronic circuit, usually containing one electron tube and associated components.

STATIC CHARGE: Charge on body either negative or positive.

STATIC ELECTRICITY: Electricity at rest as opposed to electric current.

STATOR: Stationary coils of an ac generator.

STEADY STATE: Fixed nonvarying condition.

STORAGE BATTERY: Common name for lead-acid battery used in automotive equipment.

STYLUS: Phonograph needle or jewel, which follows grooves in a record.

SUBHARMONIC: A frequency below harmonic, usually fractional part of fundamental frequency.

SUBSONIC: A frequency below the audio frequency range; infrasonic.

SUPERSONIC: Frequencies above audio frequency range.

SUPERHETERODYNE: Radio receiver in which incoming signal is converted to fixed intermediate frequency before detecting audio signal component.

SUPPRESSOR GRID: Third grid in electron tube, between screen grid and plate, to repel or suppress secondary electrons from plate.

SWEEP CIRCUIT: Periodic varying voltage applied to deflection circuits of cathode ray tube to move electron beam at linear rate.

SWITCH: Device for directing or controlling current flow in circuit.

SYNC PULSE: Abbreviation for synchronization pulse, used for triggering an oscillator or circuit.

TANK CIRCUIT: Parallel resonant circuit.

TAP: Connection made to coil at point other than its terminals.

TAPE TRANSPORT: The driving mechanism and reels of a tape recorder.

TELEVISION: Method of transmitting and receiving visual scene by radio broadcasting.

TELEVISION CHANNEL: Allocation in frequency spectrum of 6 MHz assigned to each television station for transmission of picture and sound information.

TETRODE: Electron tube with four elements including cathode, plate, control grid and screen grid.

THERMISTOR: Semiconductor device which changes resistivity with change in temperature.

THETA (Θ): Angle of rotation of vector representing selected instants at which sine wave is plotted. Angular displacement between two vectors.

THREE-PHASE ALTERNATING CURRENT: Combination of three alternating currents having their voltages displaced by 120 deg. or one-third cycle.

TICKLER: Coil used to feed back energy from output to input circuit.

TIME CONSTANT (RC): Time period required for the voltage of a capacitor in an RC circuit to increase to 63.2% of maximum value or decrease to 36.7% of maximum value.

TONE CONTROL: Adjustable filter network to emphasize either high or low frequencies in output of audio amplifier.

TRANSCEIVER: A combined transmitter and receiver.

TRANSDUCER: Device by which one form of energy may be converted to another form, such as electrical, mechanical or acoustical.

TRANSFORMER: Device which transfers energy from one circuit to another by electromagnetic induction.

TRANSFORMERS, Types: ISOLATION. Transformer with one-to-one turns ratio.

STEP-DOWN. Transformer with turns ratio greater than one. The output voltage is less than input voltage.

STEP-UP. Transformer with turns ratio of less than one. Output voltage is greater than input voltage.

TRANSIENT RESPONSE: Response to momentary signal or force.

TRANSISTOR: Semiconductor device derived from two words, transfer and resistor.

TRANSISTOR SOCKET: A small device in which the leads of a transistor are placed to provide ease in connection to circuit and to permit replacement of transistor.

TRANSMISSION LINE: Wire or wires used to conduct or guide electrical energy.

TRANSMITTER: Device for converting intelligence into electrical impulses for transmission through lines or through space from radiating antenna.

TRIMPOT: The trade name for a precision variable resistor manufactured by Bourns.

TRIODE: Three-element vacuum tube, consisting of cathode, grid and plate.

TRIVALENT: Semiconductor impurity having three valence electrons. Acceptor impurity.

TRUE POWER: Actual power absorbed in circuit.

TUNE: The process of bringing a circuit into resonance by adjusting one or more variable components.

TUNED AMPLIFIER: Amplifier employing tuned circuit for input and/or output coupling.

TUNED CIRCUIT: Circuit containing capacitance, inductance and resistance in series or parallel, which when energized at specific frequency known as its resonant frequency, an interchange of energy occurs between coil and capacitor.

TURNS RATIO: Ratio of number of turns of primary winding of transformer to number of turns of secondary winding.

TWEETER: A high frequency speaker.

UNIT MEASUREMENT: Standard increments of measurement or quantity used as means of comparison to other quantities.

UNIVERSAL MOTOR: Series ac motor which operates also on dc. Fractional horsepower ac-dc motor.

UNIVERSAL TIME CONSTANT CHART: A graph with curves representing growth and decay of voltages and currents in RC and RL circuits.

VALENCE: The chemical combining ability of an element in reference to hydrogen. The capacity of an atom to combine with other atoms to form molecules.

VECTOR: Straight line drawn to scale showing direction and magnitude of a force.

VECTOR DIAGRAM: Diagram showing direction and magnitude of several forces, such as voltage and current, resistance, reactance and impedance.

VOICE COIL: Small coil attached to speaker cone, to which signal is applied. Reaction between field of voice coil and fixed magnetic field causes mechanical movement of cone.

VOLT: A unit used to describe a difference between two levels of electrical force. This force will cause the flow of electricity between the two levels if connected by a circuit or wires. A flashlight cell will be rated at 1.5 volts. The automotive battery has a 12 volt difference between its terminals. The electricity we use at home is about 117 volts. Very often, units smaller or larger than a volt are used for convenience. Study the following terms.

MEGAVOLT equals one million volts or 1,000,000 volts.

KILOVOLT equals one thousand volts or 1000 volts.

MILLIVOLT equals one thousandth of a volt or .001 volt.

MICROVOLT equals one millionth of a volt or .000001 volt.

VOLTAGE DIVIDER: Tapped resistor or series resistors across source voltage to produce multiple voltages.

VOLTAGE DOUBLER: Rectifier circuit which produces double the input voltage.

VOLTAGE DROP: Voltage measured across resistor. voltage drop is equal to product of current times resistance in ohms. $E = IR$.

VOLTAIC CELL: Cell produced by suspending two

dissimilar elements in acid solution. Potential difference is developed by chemical action.

VOLTMETER: Meter used to measure voltage.

VOM: A common test instrument which combines a voltmeter, ohmmeter and milliammeter in one case.

VTVM: Vacuum Tube Voltmeter.

VU: Number numerically equal to number of decibels above or below reference volume level. Zero Vu represents power level of one milliwatt dissipated in 600 ohm load or voltage of .7746 volts.

WATT: This unit is a measurement of power used in an electrical circuit. You already know about mechanical power. Your car engine may be rated at one or two hundred horsepower. Power is the rate of doing work.

In electricity, work is done when an electrical current overcomes the resistance in a circuit. The result is heat which is radiated out to the surrounding air. Your electric toaster at home may be rated at 550 watts. You use 75 and 100 watt light bulbs. Other larger and smaller units are:

MEGAWATT - one million watts or 1,000,000 watts.

KILOWATT - one thousand watts or 1000 watts.

MILLIWATT - one thousandth of a watt or .001 watt.

MICROWATT - one millionth of a watt or .000001 watt.

WATT-HOUR: This term describes the use of electrical power. It is also the unit by which you purchase power from the power company. If you use one watt of power for one hour, your power consumption would be one watt-hour. This unit is very small. Therefore the meter on the outside of your home reads in KILOWATT-HOURS or KWH. If you consume one thousand watts per hour, you use one KWH. A one hundred watt lamp burning for ten hours would use one KWH. The power consumed by a circuit or device can always be found by the formula, VOLTS times AMPERES equals WATTS. Multiply your answer by time in hours and you have watt-hours.

WATTLESS POWER: Power not consumed in an ac circuit due to reactance.

WATTMETER: Meter used to measure power in watts.

WAVE: A disturbance in a medium which is a function of time and space or both. Energy may be transmitted by waves. Example: Audio and radio waves.

WAVE FORM: The shape of a wave derived from plotting its instantaneous values during a cycle against time.

WAVELENGTH: Distance between point on loop of wave to corresponding point on adjacent wave.

WHEATSTONE BRIDGE: Bridge circuit used for precision measurement of resistor.

WOOFER: A large speaker designed for reproduction of low frequency sounds. It is used in high fidelity amplifier systems.

WORK: When a force moves through a distance, work is done. It is measured in foot-pounds and Work = Force × Distance.

X-AXIS: The horizontal axis of a graph.

Y-AXIS: The vertical axis of a graph.

ZENER DIODE: A silicon diode which makes use of the breakdown properties of a PN junction. If a reverse voltage across the diode is progressively increased, a point will be reached when the current will greatly increase beyond its normal cutoff value. This voltage point is called the Zener voltage.

ZENER VOLTAGE: The reverse voltage at which the breakdown occurs in a Zener diode.

INDEX